Help for the Technology Shy

A COMMONSENSE APPROACH TO SAFE, PRODUCTIVE, STRESS-FREE TECHNOLOGY USE

— **Part I** —

MICHAEL GORZKA

Welcome!

Help for the Technology Shy - Part I

MICHAEL GORZKA

"Success is a journey not a destination..."

Published by Squirrels Drinking Coffee

Help for the Technology Shy - Part I

International Standard Book Number: 978-1-7349815-5-1

Table of Contents

Michael Gorzka
Help for the Technology Shy.
Part II

Why Do You Need a Technology Foundation?

In a nutshell, for the same reason you would want a solid foundation underneath your house.

Italics and Use of icons

Geeky technology terms and other key concepts will be *italicized* and will be cross-referenced in this book's index (and are Help for the Technology Shy blog topics).

Due to space considerations, the real-life examples & case studies in this book will be "painted in broad strokes."

More details on the examples in this book (and other particularly good stuff besides) can be found within this book's accompanying website.

And this book's accompanying website is located at www.technologyshy.com.

Query: "I'm technology **shy**, O, Tech Wizard! That's why I'm reading this book! How am I supposed to get to your no doubt fabulous and transcendent accompanying website?"

Watch video demonstrations for what we cover in this book (and more!)

Answer: We will cover how to search the web and visit websites in this book.

You can also visit your local library or adult community center and ask a friendly staff member to help you get to www.technologyshy.com.

Ask for headphones so you can hear the melodious sounds of your Tech Wizard author's voice.

This stylish "finger wearing wizard hat" illustration is a reminder icon.

The nuts & bolts, the nitty gritty technical details will **vary** between devices but the *technology basics* will be just the same.

A memo from the Help for the Technology Shy "Tough Love Department"...❣️

'Nuff said!

This note-taking squirrel (or a chipmunk 🐿 at the end of a passage) is shorthand for "In a nutshell..." and is also a reminder to document your technology use.

Movie-related illustrations & emojis are reminders that this book is supplemented by videos and an instructional DVD.

A reminder that technology can be a bit (or a lot) mystifying at times, but it is definitely not rocket science.

This cute fish (which looks a bit like "Mr. Limpet") is a gentle reminder that one of this book's primary goals is to teach you how to fish instead of simply handing you a fish – as a well as a message about believing in one's self and abilities.

The beginning is not only a great place to start, it's a great place to regroup (with a nice cup of tea).
— Tech Wizard Mike

Tea cup and tea brewing illustrations are reminders for when you feel discouraged / dissatisfied / distressed / disenchanted / disappointed with a particular technology application (or with technology in general) to…

1. *stop* ✋
2. *brew a nice cup of tea* ☕
3. relax, regroup, *breathe*
4. and then start again from the beginning – when you are ready to do so

Tea and coffee cups are also spiffy reminders to apply the same logical procedures and visualized outcomes you use to *brew a cup of tea* and/or make coffee to your technology use.

You should have your technology notebook within easy reach during technology use in order to record and/or retrieve any relevant to the task at hand information 📒

"gone the way of the dodo"

Whenever you see a dodo bird in this book, it will signal that the specific example we are using may be **out of date** by the time you read this – but the process of thoughtfully applying the *timeless technology basics* (which we will cover in this book and *supporting materials*) will be just the same.

Menus (especially *context menus*) are the most often overlooked tools for getting stuff done with technology.

This is the "Info" icon – it marks passages that contain (at the risk of sounding immodest) your Tech Wizard Tour Guide's unique insight and perspectives.

Just like in real life, it is **very** important to pay close attention to *visual cues* during technology use

If you see a "thumb" illustration near a particular passage, you can consider it to be shorthand for "In general, follow these guidelines…" or "Take a course of action like this…"

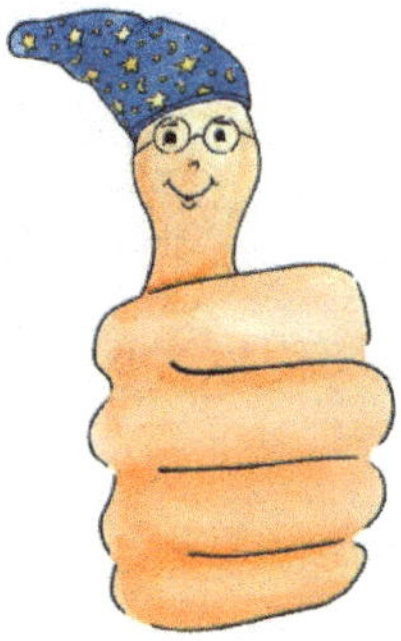

This signifies a key concept – something you will especially need in order to achieve productive, stress-free technology use

Context matters! As much as in the technology world as it does in the real world 🏝️

Simply *noodling around* devices, apps and websites (with no particular objective or expectations of yourself) can be extraordinarily beneficial 🍜

Use only what is necessary and "KISS" your technology use 💋

(but not this way!)

Your normally unobtrusive Tech Wizard author does indulge in the occasional soapboxing…

Chapter 1: Introduction

Who this book is for...

If you are somewhat (or totally) mystified by modern gadgetry, you have come to the right place.

If your device is not working as it should, you have come to the right place.

If you have in your possession, a computer, tablet, wearable, smartphone, whatsis or thingamajig and would like to **do something with it** but the thought of powering it on gives you the whim-whams, you have come to the right place!

"It's an entirely different world and I don't understand it, one bit"

If this sounds like you, then you have come to the right place!

If you feel **less than comfortable and proficient** with whichever technologies you have to use for your job or your academics, you have come to the right place.

If you would like to **explore** the possibilities of using one or more technologies to enrich your life in some way but don't quite know where to begin, then you have come to the right place.

If you have virtually **no interest** in 21st century technologies but have them to be a necessity (or at least very helpful) in order to independently function in the modern world, you have come to the right place!

What are the goals of this book?

Glad you asked! The goals of this book are twofold...

First of all, your Tech Wizard scribe / new best friend wants to put you **at ease** with technology.

He wants to initiate a rapprochement between you and the overabundance of devices we have access to **and** the near limitless number of things we can **do** with those devices…

…as many of your author's technology shy friends have had a "combative," shall we say, relationship with technology which often leads to less than optimal results…

Goal #2

Our second goal is to enable you to comfortably **get stuff done** utilizing the *devices / applications / cloud services* of your choosing.

How is this book different?

My goodness, your stalwart & intrepid Tech Wizard author **and** spiffy librarian has heard that statement **a lot** over the years!

(And the fact that technology help books have a shorter shelf life than milk is only one reason.)

There are a **plethora** of computers, tablets, smartphones, audio visual contraptions and various gadgets of every stripe.

And there is a **plethora** of things you can do with any of those shiny gadgets.

And there is a **plethora** of things that can go awry if you do not use said gadgets with thought and intention.

NOTE: From here on, we will use the word "device" as a **catch-all term** (a metaphorical "dumping ground" as it were 🤭) for computers, tablets, smartphones, wearables, A/V contraptions, *mystery gadgets*, et al.

People are often dissatisfied (to say the least) by the technology help books they pick off library and book store shelves because...

1. They have a **different** device than the one that the book covers.

2. They have a newer or older **model** of the device that the book covers.

3. They might be using an "obsolete" device or a device that is infected with *malware* 🦠

4. The screens on their devices look different from the ones shown in the book.

5. The **specific thing** they want to do with their device of choice is not covered in the book.

Due to the **constantly** evolving (or devolving, depending on your point of view) nature of technology...

...putting step-by-step instructions for **specific tasks and projects** into a printed book would be wildly impractical because that book would be out of date before it was even printed.

Therefore the only practical solution is to teach you how to fish – instead of simply handing you a fish.

Teaching you how to fish!

Apps and **services** come and go, devices turn into pumpkins and **appearances** frequently change – and vary widely amongst devices.

But a mouse click has been a mouse click since 1985 or so and a screen tap has been a screen tap since circa 2007. And we're looking at the year 1867 for keyboard presses 😯

These three easy peasy lemon squeezy timeless technology basics will be present and accounted for in your *Getting Stuff Done with Technology toolbox:*

Which computers, tablets, smartphones, wearables, mystery gadgets, applications and online services will you cover?

Actually all of them! 

And none in particular. 

Instead of focusing on one specific type of device or application or online service, this book has been written to **cover all the bases** so to speak. 

So regardless of what type of device you have and what you might want to actually **do** with said device – this book is here to help you.

"The more things change, the more they stay the same."

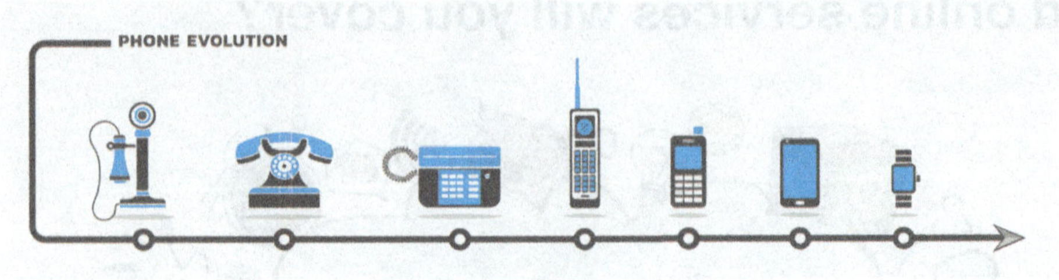

This phrase is often said in a resigned or ironic tone but when it comes to technology use, it can be your rock 🪨, your anchor ⚓ and your mantra ✌️

As previously mentioned, technology changes **often** (sometimes from moment to moment 😮).

But the good news is, the thought, organization, logical procedures and applied *focus* necessary for getting stuff done with technology have not changed at all – and you have found this book 🙂

Meet your new best friend, your friendly technology tour guide & nonjudgmental muse

Michael Gorzka (c'est moi) is a retired public librarian and has been teaching adults how to properly use computers and other forms of technology since the mid-1990s (and also prefers to speak of himself in the third person).

"It was an enormous beast of a thing..."

This is how your Tech Wizard scribbler remembers the first computer he ever used way back in 1981. He found it to be **very** arcane and intimidating!

Your author's first assignment was to get this "big iron" to print his name by feeding punch cards into it – It took him a week!

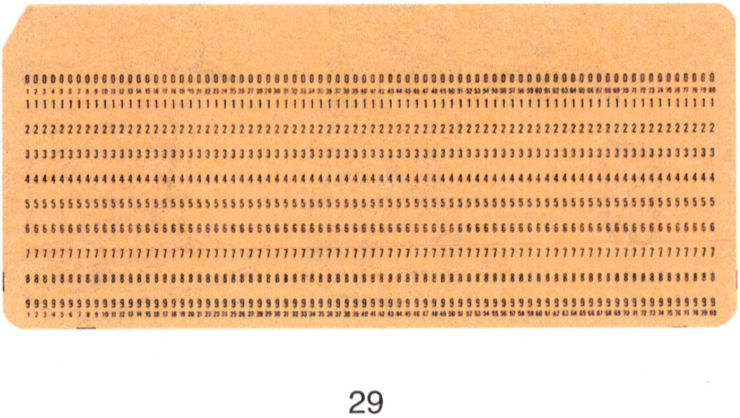

After making it through that class (thanks to some *good help* from his teacher), your author swore up and down he would never touch another computer again. What good were they?

If you had told him at that time that he would one day be a snazzy "Tech Wizard" and able to resolve many plights related to the use (and misguided use) of technology – and would even eventually pen a "Help for the Technology Shy" book, he would have looked at you as if you had grown a second head!

If a former technophobe like myself can become comfortable, proficient and productive with technology, you can too!

Let's begin!

"Be willing to be a beginner every single morning."

— Meister Eckhart

The analogy of the snake and the rope

A man walks at night along a path. He sees a poisonous snake barring his way and turns and runs in the opposite direction.

As he returns along the same path in the morning, he finds a coiled rope on the ground.

He realizes that in the darkness, he mistook the coiled rope as a snake and it dawns on him, in the dark it is hard to see reality as it truly is. In the light of day, we see more clearly.

In your Tech Wizard friend's experience, many people are unable to make positive, productive use of technology because they are seeing computers, tablets, smartphones and other forms of 21st century gadgetry as deadly serpents instead of merely coils of rope.

This misidentification can be subdued, overpowered, outdone, thwarted and vanquished through **mental** and **emotional preparation** for contemporary technology use as ably outlined in subsequent chapters in this book and *supplemental materials*.

Seven ways to reduce your technology anxiety

1. Breathe

Breathe.

BREATHE

BREATH IN.

BREATHE OUT

BREATHE DEEP.

BREATH SLOW

Comfortable, deep breathing is the key to relaxation. All the traditional relaxation methods (yoga, meditation, hypnosis) place a central emphasis on breathing.

If you begin to feel stressed, nervous, apprehensive, anxious (or all of the above) during technology use, try some breathing exercises.

2. Approach technology gradually...

The Crow Pose (Bakasana) is one of the first arm balances learned by yoga students.

In all variations, Crow is an arm balancing posture in which hands are planted on the floor, shins rest upon upper arms, and feet lift up.

You can eventually get into this pose through patience, persistence and practice (aka "the 3 P's").

You can become comfortable and proficient with technology the same way.

Just be prepared to fall on your face a few times (figuratively speaking of course).

3. Write down (or sketch) your current objective, your *visualized outcome*, what you wish to use technology to accomplish – in a way that makes sense to you.

This will make your objective a more tangible thing.

It will help keep you grounded and outcome focused – which in turn will help you feel more relaxed and in control.

You can simply refer to your written or sketched objective if you feel you have lost your way or have become **distracted** by things that have nothing to do with your current goal / objective / target / *visualized outcome* 🙌

Conceptual understanding of your objective will be much more beneficial than reliance on *step-by-step instructions*.

4. Seek out positive & productive collaborations.

Storage versus memory

5. Keep in mind that nothing is engraved in stone. Nothing will be broken that can't be fixed.

If you make a faux pas or take a wrong turn during technology use, please keep in mind that in the grand scheme of things, it will be of no consequence. It will be nothing even **remotely resembling** biggie fries.

That being said, if you even **suspect** your device has *malware* or if you see unexpected financial charges, then get *good help* straight away.

6. I love coffee, I love tea
I love the java jive and it loves me

Know that a cup of tea will always be there if you need it.

In addition to curbing your stress and anxiety (plus a host of other benefits), making a nice cup of tea involves thought, selection of options, taking logical steps, a calm demeanor and a *visualized outcome* – all of which are **indispensable** for productive, stress-free technology use.

7. Consider any technology-related *objective* to be a jigsaw puzzle
And what do you do when you are putting together a jigsaw puzzle? You focus on **one piece at a time**.

It would not make sense to try to fit all the pieces together simultaneously. Approach your technology use the same way.

For example, *focus* on sending messages using an *app* on your device (which you may have came across by *noodling around*) **until** you know the procedure by heart – **until** you feel comfortable and proficient with that particular task.

Use *visual cues* to help put your "puzzle" together, piece-by-piece...

Bonus tip: When you find yourself in a state of agitation, when you find yourself "white knuckling" your device, stop and take a break!

All of the various and sundry 21st century devices, applications, services and resources will still be here when you get back.

You are <u>not</u> a dinosaur 🦖

Many of my technology shy friends have called themselves "dinosaurs" due to their perceived lack of technology expertise.

This cannot possibly be true because dinosaurs are extinct.

If you are reading this then you are alive and so cannot be a dinosaur – quod erat demonstrandum.

Not You!
(not even on a bad day)

Hey, it's easy to feel comfortable with technology if you were **born** into it.

Case in point

But what if you are coming fashionably late to the technology party?

(But truth to tell, it's not much of a party…)

Whiz Kids?

One of my technology shy friends emphatically stated that children are able to use technology so easily because "they are so smart!"

You can see children who are preschool age **effortlessly** using all kinds of gadgets.

How can these kiddos (who cannot even read & write yet) be smarter than we are?

The obvious answer is **they are not**.

Children – due to their total lack of life experiences and ego – are able to approach technology with a **blank slate**.

They have no misconceptions or apprehensions about contemporary technologies and so are able to freely use them in a stress-free manner.

And you can too!

But in order to do that, you first need to press your big ol' emotional reset button:

Okay, so you may have made a lot of missteps, slip-ups, bloopers, flubs, miscalculations and various wrong turns with technology.

And they all add up to "so what⁉"

If you look at the stern of a boat you are traveling the high seas on, you will see an area of displaced water known as "wake" 🌊

This wake is only a **record** of where you have been on the boat, it does not control where you are **going**.

So for heavens sake, don't beat yourself up (or your devices).

And there is certainly no need to hold them personally accountable...

Now without further adieu, we are going to wipe your technology slate clean:

There! You can bid all of your technology-related mishaps, miscues and setbacks a fond fondue because they are gone!

They cannot drag you down anymore 👍

You're welcome!

Now, with your "inner dinosaur emotional baggage" thus thrown to the metaphorical curb…

… you can now move **forward**.

Be simple, don't carry the baggage of the past, open your hands, and let it go.

— Dr. Debasish Mridha

The most important thing to know about technology is…

…our state of the art, high-tech, shiny devices are all as dumb as a bag of hammers...

...or a box of rocks:

Our devices only know what we tell them.

Therefore we must be **very** careful what we tell them.

We can tell our devices to **do things** by (depending on the device):

- clicking the mouse
- pressing keyboard keys
- pushing buttons
- tapping the screen
- using verbal commands 🗣

Just like in real life, in the technology world, actions often have reactions.

To avoid unexpected and adverse consequences, it is so very important to proceed **mindfully** during technology use so that we do not **inadvertently** tell our devices to do something we do not want them to do – such as install a *rogue application*.

Our "input devices" should not be the least bit frivolous!

"**GIGO**" stands for garbage in, garbage out and is used to express the idea that in computing and other spheres, incorrect or poor quality input will always produce faulty output.

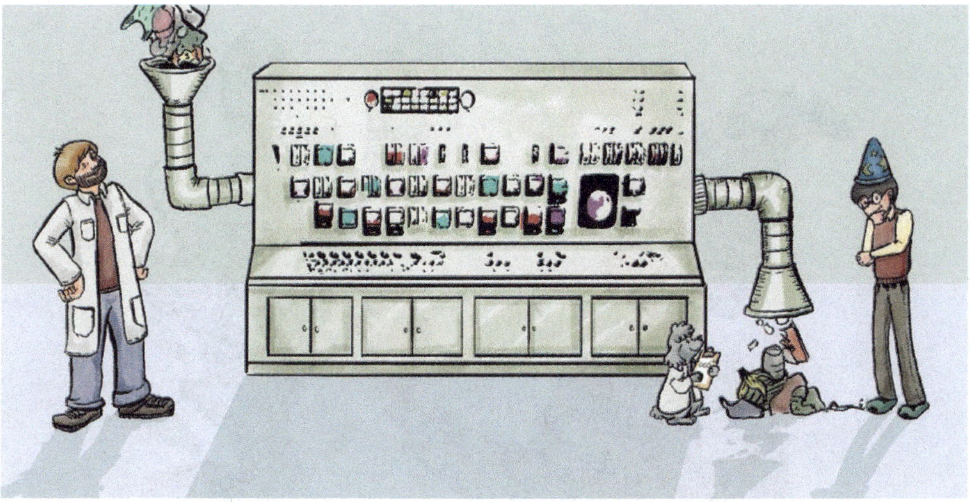

In other words, you are going to get out of technology exactly what you put into to it.

The same methods and logical procedures that we apply to traditional household appliances **also** apply to computers, smartphones, tablets and other shiny gadgets.

For obvious reasons 😉

My friend, using **any** device that can connect to the Internet is most definitely **not** like watching a summer blockbuster at your local multiplex.

If you approach any computer, tablet, smartphone, wearable, whatsis, et al. with a lack of proper thought and organization combined with unrealistic expectations, your end result has a high risk of being...

...less than favorable.

Productive, stress-free technology use begins, continues and ends with what you are thinking.

The most expensive, the most advanced, the most shiny, the most high-tech gadget on the planet will be your blunt instrument for getting stuff done and nothing more.

You don't have to be a smart cookie to put any shiny gadget to productive use – but simply a calm, cool and organized cookie.

And your author will dutifully pontificate on how to become such a cookie 🍪

Twenty-first century technologies are flawed (they pretty much have to be)

This is due to the fact that our devices, apps and any service that you can find online was created by **humans** who are themselves flawed and who are often working under adverse conditions.

Yes, your Tech Wizard friend has been here himself!

We need to expect the unexpected and accept the many imperfections, pitfalls, stumbling blocks and necessary detours that will come our way during everyday technology use.

Making productive use of 21st century technologies is not rocket science nor is it an exact science which brings us to the very next section.

Let's Talk Rocket Science! 🚀 (Or Not 🙂)

Please excuse the false advertising but we are not going **anywhere** in this chapter ✨

Just the opposite in fact – while using 21st century technologies, we need to keep our feet (and hearts & minds) firmly planted on terra firma 🌱

There is no reason to leave Mother Earth 🌎 because my friend, we do **not** need to have the IQ of a Rocket Scientist to make productive use of modern technologies (and it is probably best that we don't 😉) .

The necessary background, knowledge and skillset we need in order to effectively use technology is known collectively as life💓

Case in point: Gather what we need 🎒, proceed at a methodical pace, stay calm and focused while being conscious of our surrounds

The dickens, you say!?!

Many people rail against technology because it is so "complicated" and "arcane" 🤯

But let's get real, we're talking about such things as pushing buttons, clicking a mouse, tapping screens, pressing keys on a keyboard and filling out forms here – and calling such operations "complex" would be a bit of a stretch.

" When you put it like that, it makes complete sense."

Your blueprint to achieve productive, stress-free technology use should not look like this...

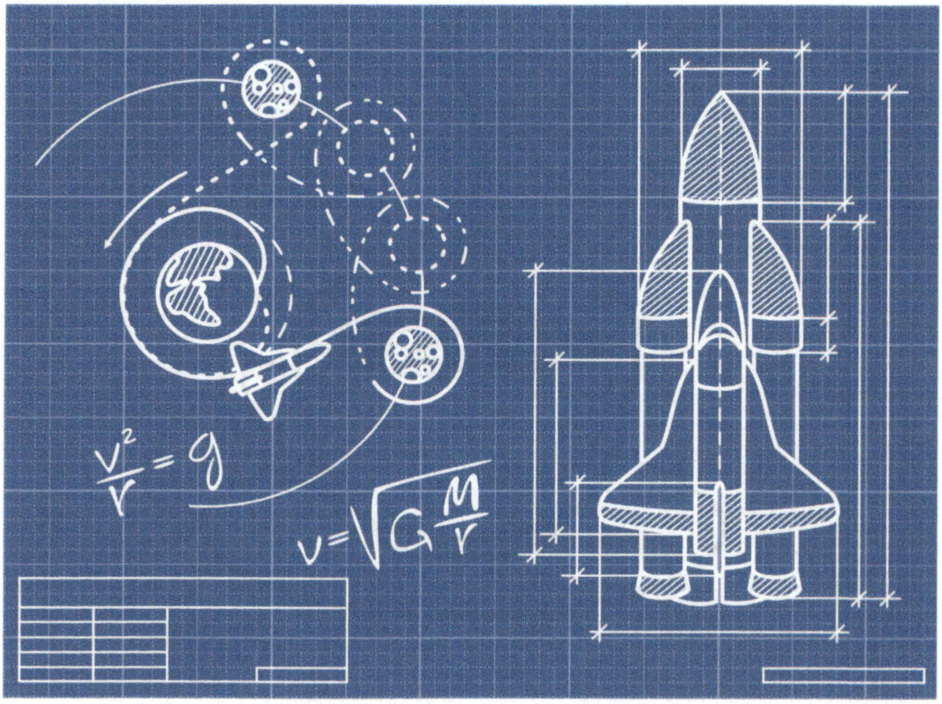

...but more like this...

How to make the perfect cup of tea

(Yes, this is a technology help book but please bear with me here.)

1. Start with fresh, cold water...
2. Place a tea bag in your favorite cup or mug.
3. Bring water to a rolling boil and immediately pour over your tea bag.
4. Steep for a good 3 to 5 minutes...
5. Remove the tea bag, relax and enjoy!

Question: What does making a perfect cup of tea have to do with becoming comfortable and proficient with technology?

Answer: Everything, really.

When you make a cup of tea you have a desired goal, an **objective**, a specific outcome in mind.

And to realize this objective, you methodically and mindfully follow a sequence of well-reasoned steps.

For example, you cannot pour the tea, add whatever you like to it and **then** heat the water – it wouldn't make logical sense to do so 🤔

Successful tea brewing also involves **patience** (3 - 5 minutes for steeping) and realistic expectations (e.g. your tea will not be **hot** unless you are willing to heat some water – however which way you want to do it ♨️).

Please approach **getting stuff done with technology** as you would brew a cup of tea.

Carry out each step in the process with thought, intention and understanding – and with a specific outcome in mind 🙌

If you do not understand **what** you are doing and/or **why** you are doing something then **stop** ✋; make a cup of tea 🍵 and start again from the beginning (when you are ready to do so).

There is something that you want to do...

You have an **idea**... 💡

Somebody gives you an *objective*...

There is something you have a hankering to do...

An impetus of some sort arrives prompting you to take action...

(“Samantha” makes her own yoga clothes...)

...this leads to a **visualization** of your desired outcome:

Next you can **think about** 🤔 / **find out** 🔎 💬 / **sketch out** ✏️ what is needed to achieve that outcome.

For example, Samantha is going to create a presentation for investors which will include...

- artist drawings of her planned new outfits
- descriptions of her designer yoga outfits (e.g. dyes, textures & fabrics)
- photos of women wearing her yoga clothes
- testimonials from existing customers
- videos showing how comfortable & functional Sam's clothes are during yoga practice

1. with a thorough understanding of what she wants to do, Sam will then mindfully implement what is needed using suitable *device*(s), *app*(s) and/or *cloud service*(s)...

2. getting *quality help* if she needs it...

3. not losing sight of the forest thru the trees...

"Hold on, where's the forest again?"

4. returning to the drawing board whenever necessary...

Remember the time-honored *GIGO* principle and the fact that our devices are all as *dumb as bag of hammers*. Your success begins, continues and ends with what is inside your head.

Chapter 3: Your Technology Foundation

Good Grief! Look at All This Stuff! 🤯

Geez, Louise!

If you are feeling overwhelmed and disheartened, rest assured you are in very good company.

And the still more good news is, having clear *objectives* can make a very positive difference…

…and help you avoid losing sight of the forest through the trees.

KISS = "Keep It Simple, Snookums"

But not this way!
(and do not forget about the things that really matter!)

The more devices, apps, resources and services we have available to us, the more important it is to KISS our technology use.

Start with your technology-related *objectives* and **then** seek out only the *devices*, *peripherals*, *apps* and *cloud services* you need to realize those objectives.

With that being said, take a peek at these...

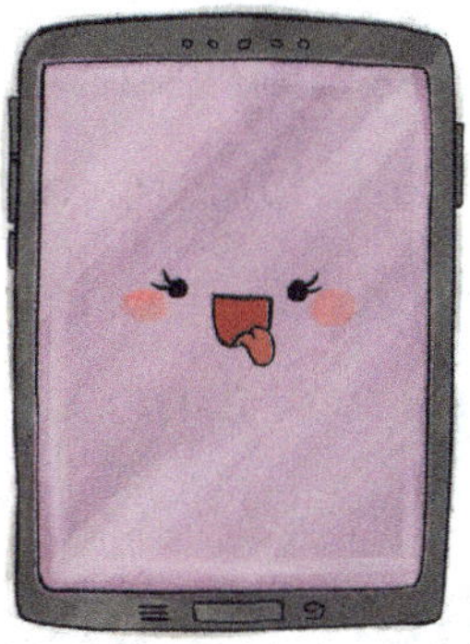

So yes to be sure, there are computers, there are smartphones, there are tablets, there are audio-visual boxes, there are wearables, there are e-book readers and of course there are *mystery gadgets* (which we will get to shortly).

These are simply some of the **steadily evolving** device options we have available to us with many more on the horizon and we are free to simply disregard that which we have no compelling use for.

Storage options

In addition to our device(s) of choice, there are **a lot** of things (physical & non-physical) we can store our "stuff" (e.g. photos, videos, documents, music, et al.) on or in for safekeeping:

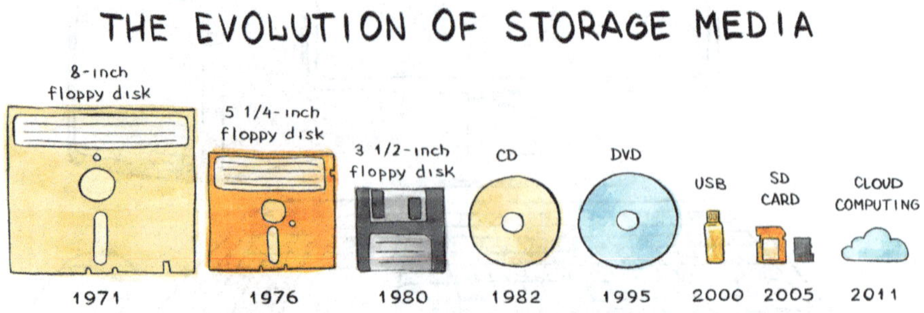

Peripherals

There are an abundance of *peripherals* we can connect (in some way shape or form) to various devices such as:

- cameras
- external storage media
- headphones
- keyboards
- "mice"
- microphones
- printers
- scanners
- speakers

"Cloud services"

And yes to be sure, we have a profusion of cloud-based services.

Cloud services are a boon for people who cannot afford their own devices (or simply wish to not own their own devices) because they enable us to create and **save** documents, spreadsheets, presentations (and more!) through many *devices* that can connect to the Internet 🍿

Icons

If your device has a screen, on it you will most likely see small but colorful pictures called *icons* which will represent **shortcuts** to various things on your device – including *files*, *folders*, *applications* (or "apps") and *utilities*.

The appearance and **location** of icons will vary between devices.

And icons will also depend on **you** because many icons can be manually added, removed and rearranged.

Some icons will have *visual cues* 👀 (to visually convey such things as "hey, you have a new message" and "you have previously opened this"):

Interfaces

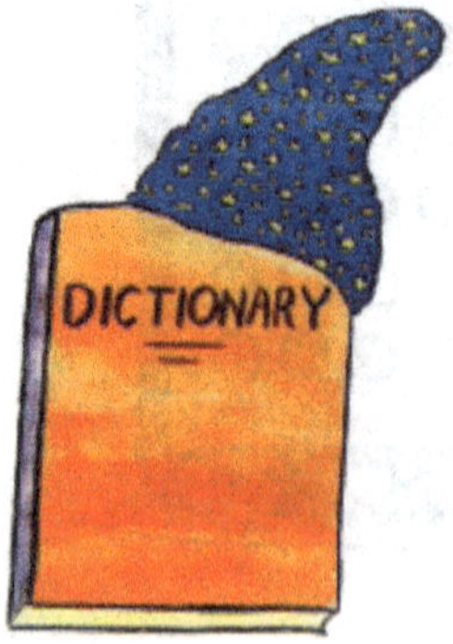

in·ter·face

Noun

the place where two systems come together and have an effect on each other, or a connection between two computers or between a person and a device

Rule of thumb: When you are confronted with any new interface – on any device – simply, pause, *breathe*, put on your fuzzy slippers and make a cup of tea (at least mentally) and scrutinize the interface **before** you start interacting with it.

We need to be calm, observant and visual during technology use 🚦

If you get lost, look for *breadcrumbs* (e.g. "back" and "home" buttons) to guide you.

Files

"Files" are anything you can create with your device or save to your device (or access through a *cloud*) including photos, documents, spreadsheets and presentations.

In addition to file names, files will have *file name extensions* such as resume.**doc** and inventory.**xls**.

A *file name extension* will tell your device what the file **is**, exactly.

Apps

Applications (or "apps") enable us to **do things** with our devices such as...

- look things up
- check the weather
- send messages
- play music
- create, format & save documents / presentations / spreadsheets
- and many more

(The appearance of app icons will vary widely between devices.)

Utilities

A utility is system software intended to analyze, configure, monitor, or help maintain a computer. A utility may be included with an operating system or installed separately.

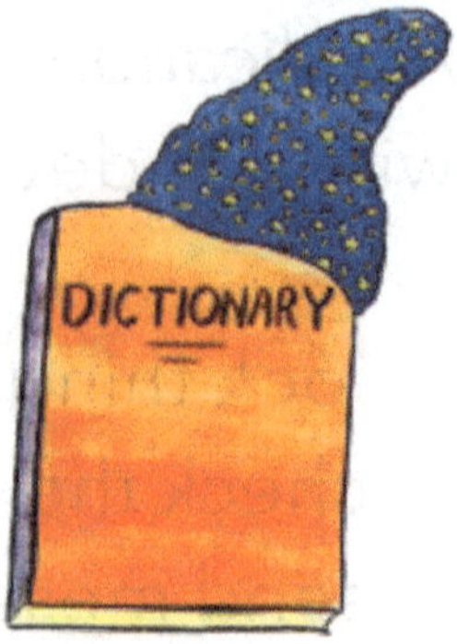

We couldn't think of a representative image for utilities so here are Matt & Ziggy rescuing me from the alien zoo. This story is at www.technologyshy.com/matt.

Operating systems

An operating system (or "OS") is the brains of your device, it determines (amongst other things) where things on your device will be, how things on your device will look and how things on your device will work (including *keyboard shortcuts*).

Mystery gadgets

Every so often, somebody will bring in a device, apparatus, contraption, thingamajig, or gizmo that your Tech Wizard chronicler has never seen before.

These devices are routinely cheap knockoffs made from disreputable or fly-by-night manufacturers with brand names that nobody has ever heard of.

In addition to not functioning properly, a cheap knockoff may in fact be **dangerous** to use to access the Internet due to the fact that such a device would have subpar security – making it (in the immortal words of Bugs Bunny) "a wolf in cheap clothing."

How do I know if I am the not-so-proud owner of a cheap knockoff?

Answer: Here's one way to find out, examine its external case for a brand name and take it to your local library and see if you can find it in Consumer Reports.

If the device is not to be found in Consumer Reports (Hello, Librarian!), you can consider it to be a cheap knockoff and your Tech Wizard author's recommendation is to return it (if possible).

Counterfeit devices

Bogus, sham computers were probably not much of an issue in 1981…

Unfortunately it is much easier to create fakes of 21st century devices – and not surprisingly, counterfeiters try to create faux versions of the more expensive, high-end devices.

If your device is not working properly and/or comes in shoddy packaging, return it 🐿

Reputable mystery gadgets

There are also reputable mystery gadgets that are either new to the market and/or not part of the mainstream technology market.

Thoughtful analysis 🤔

These are niche products that often fulfill a specific, specialized function.

(Although it is sometimes less than clear what that specific function might be.)

If you have such a device, try to identify it (as previously described) and then try *searching the web* for more information on it – including its manual:

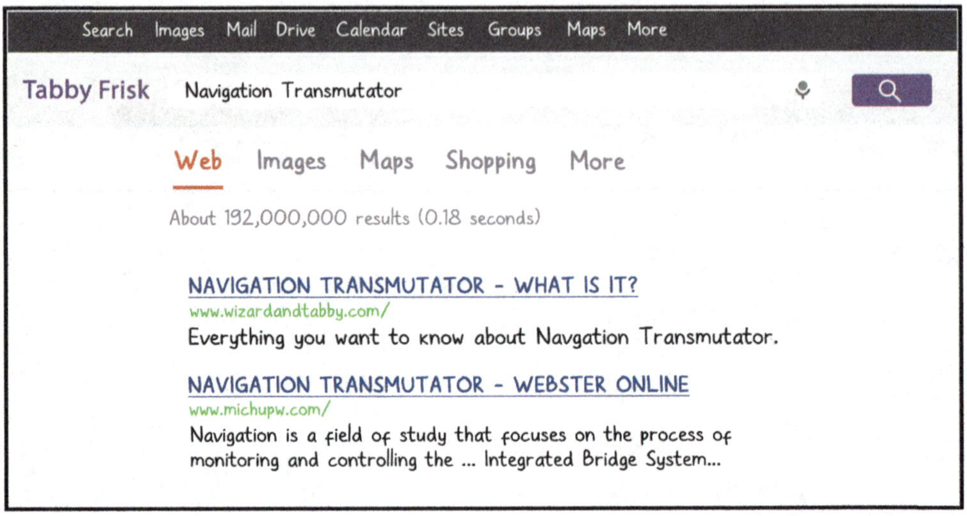

Web search results page

Rule of thumb: Whenever you are confronted with a new *device*…

1. inspect it carefully 👀
2. if necessary, search the web for info on it 🔎
3. *noodle around* with it (if possible)
4. understand what it is and what it can do 🤔
5. document your findings / write down your thoughts 📒

Building Your Technology Foundation

You can consider your emotional and mental preparation to be the **bedrock** on which your technology foundation is resting on.

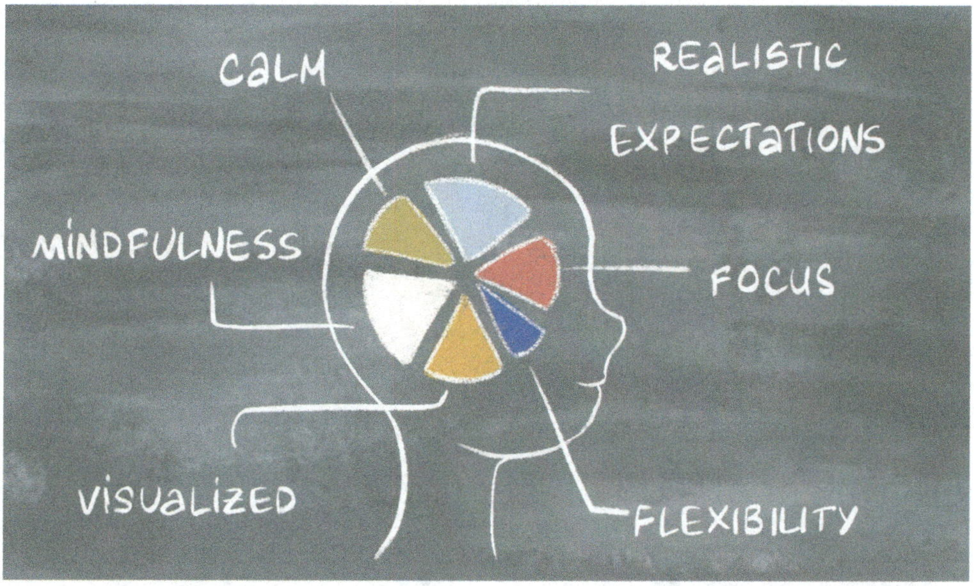

In a nutshell 🐿️, if you do not have the aforementioned emotional and mental "bedrock" **and** a sturdy, rock solid technology foundation placed on top of it...

...if in fact you do not have all of your technology-related ducks in a row and facing in the right direction...

... then your technology use will be the equivalent of a dumpster fire 🚒

Think of it this way, you would not want to worry about changing the woodwork in your dining room while the foundation of your house is wobbling underneath you.

Nor would you want to make serious use of technology without a solid foundation underneath you:

Since the mid-1990s, your author has seen **many** people try to do just that and it hasn't been pretty!

A single missing "brick" in your technology foundation can cause your technology house to collapse into a heap – and cause you a great deal of consternation and stress – faster than you can say Jack Robinson.

Take control of your technology use

After your technology foundation is in place, you can *focus* on problem solving and **getting stuff done** ✅

To quickly summarize...

You need to know the **basic** geeky terms which are covered throughout this book and *indexed*.

You need to know the *technology basics* (which are surprisingly few in number).

You need to know what kind of device you are using and its *operating system*.

You need to have an organizational system in place to keep track of your passwords.

And you need to ensure that your device is free from *malware*.

Yes, this may sound overwhelming to you but fret not because you have found this book and over the next few chapters, we are going to **work together** to create your technology foundation from each of those building blocks.

The three "timeless" technology basics

The technology basics that we will briefly discuss here are easy peasy lemon squeezy to learn – but they are also very **powerful tools** which must be used with thought and intention while always being well-aware of your *context*.

Although technology is evolving even as we speak…

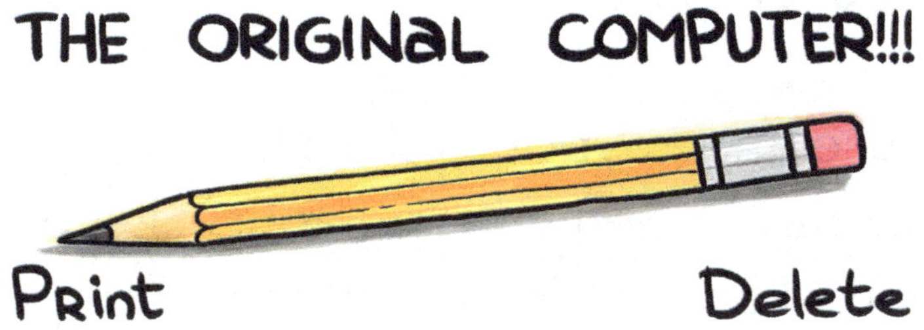

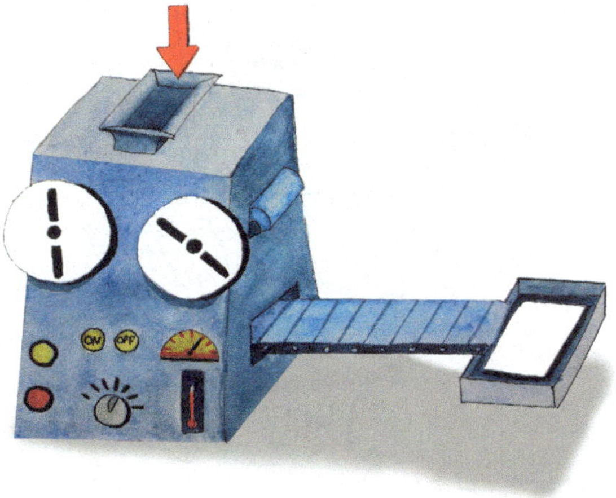

… the three **basic** technology skills that we will discuss in this chapter have stayed surprisingly constant over the years.

The mouse and keyboard skills that we discuss in this book have been in popular use since the introduction of the first Macintosh® computer.

The touch gestures have been the same since 2007 (or so).

And so, submitted for your perusal, is a compendious accounting of the three (count 'em, three) "timeless" technology basics:

1. point and click the mouse 🐭

2. tap the screen, swipe the screen, pinch the screen (ouch! 😧)

3. press keys on the keyboard

And that's it! 🎉 🎈 🎊

For space considerations, we are only going to briefly cover each of the technology basics in this chapter and provide a few insights about each one.

If you feel you need more instruction on any of these technology basics, please see the "Where to Go From Here" chapter.

Gone the way of the dodo

The traditional mouse may no longer be in popular use by the time you read this.

But even if this is the case, your author implores you, beseeches you, entreats you, **importunes** you to read this section anyway as the things we will talk about in relation to "mice" will apply to not only touch screen devices (and trackpads) but technology use as a whole.

Technology basic 1: Introducing… the mouse!

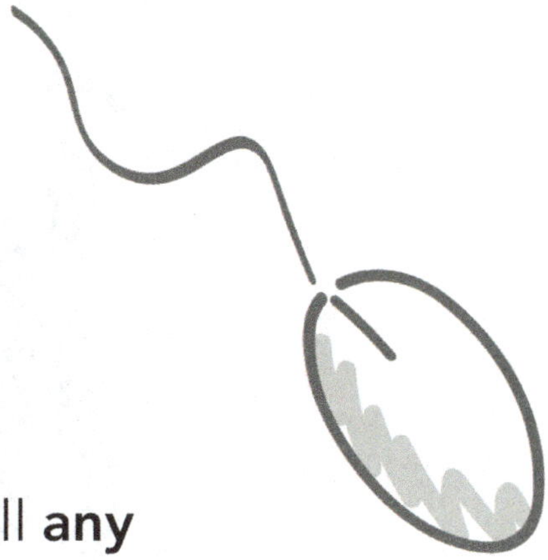

It is of utmost importance to keep in mind that the mouse (regardless of its physical appearance) is used for **input**.

You can use the mouse to tell **any device that supports a mouse** what you want it to do.

Depending on your device, you can use a mouse to…

- move the *screen arrow* or "cursor" (admittedly not very exciting on its own 😴)
- navigate a device (or simply explore or "*noodle around*" it)
- cause *menus* to appear (and then in turn carry out a great number of tasks from those menus)
- open *files* and *folders*
- open applications (or "*apps*"), close applications, switch between open applications
- perform basic operations such as restarting a device, shutting it down and putting it into sleep mode 🍿
- and more!

The screen arrow

The screen arrow is a point of *focus*. You need to know where the screen arrow is – and what it is over – **before** you click the mouse.

Your Tech Wizard author has seen many people haphazardly, almost reflexively click things on the screen as if they were playing pinball – and get lost in the weeds as a result.

This is no way to run a railroad. Clicking the mouse willy-nilly often results in confusion, lost work and **Trouble** (with a capital "T") 😳

Question: How can one avoid confusion, lost work and Trouble with a with a capital "T"?

Answer: Proceed mindfully. Use the mouse with thought and intention.

Keep in mind that *context* will always be a factor when you are using the mouse.

The screen arrow will change **depending** on what the tip of the screen arrow is currently over.

(The current state of the screen arrow is a *visual cue* 👀)

For example, when the screen arrow is **anywhere** over a text box, it will change into an i-beam pointer:

The mouse is very easy to use and it is also very easy to **misuse**.

Don’t be a sucker! Beware of things like this! Use common sense❣

The mouse under your command can be a diligent worker or a no-account, tipsy miscreant – It’s up to you.

For example, "Sally" – who had been filling out an online job application using one the library's public computers – called me over in a state of near panic.

When your author asked Sally what had occurred, she said:

"I don't know. I went click click click and I don't know what happened!"

Sally had inadvertently closed her web browser window by clicking that window's **close** button – causing her to lose all of her work.

Sally did not know that the **tip** of her screen arrow was currently over the window close ("**X**") button when she clicked the mouse.

We need to apply what we have learned from living in the real world to the technology world.

For example, if you are traveling through a strange neighborhood (or even a familiar neighborhood), you need to pay close attention to *visual cues* which include other automobiles, road signs, pedestrians, bicyclists, and traffic lights in order to safely reach your destination.

(If you pay close attention to visual cues, you will often not need step-by-step instructions)

If you do not pay close attention to *visual cues*, there is a huge risk of either getting lost or incurring an accident.

We need to approach our technology use the same way.

Remember our shiny devices are all as *dumb as bag of hammers* – which is why your **brain** is in your *Getting Stuff Done With Technology toolbox*.

All the other tools are **useless** without it ❣

Question: How will **my** mouse work?

Answer: How you can use your mouse will depend on what kind of a mouse you have, the type and brand of device that you are using, and perhaps how your mouse has been configured to work through your device's *Settings.*

This is good to know because the built-in mouse on laptop or notebook computers (called "trackpads") may or may not have *buttons*.

The trackpad **itself** could be one large button:

If you move the mouse with your hand or your finger across your device's trackpad or surface of the mouse (again depending on *which device* & **which mouse** you are using) the screen arrow should move.

Simply moving the screen arrow around seemingly does nothing but when you do so, you are moving **focus** – which can affect the outcome of your mouse click.

As in real life, **cause-and-effect** will always be present during technology use.

Every time you click the mouse (or touch the screen on a touch-enabled device)...

- You are sending the device a **signal**.

- You are giving it a **command**.

- You are telling the device you want it to **do something**.

Technology basic #2: Touch screen gestures

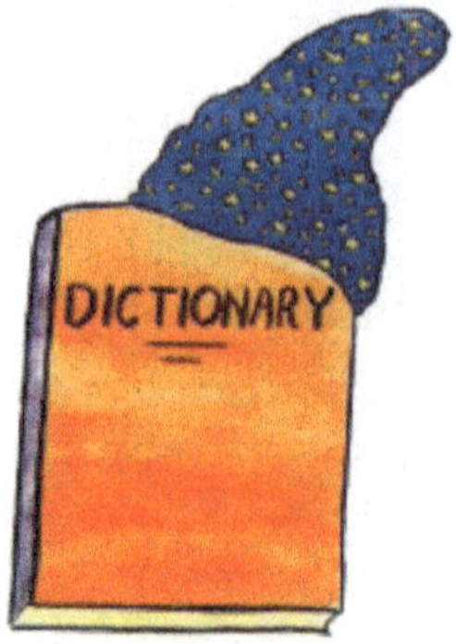

Touch screen noun
\ ˈtəch-ˌskrēn\

: a display screen on which the user selects options (as from a menu) by touching the screen

A touch screen is an input device 🐿️

A user **interacts** with the device by touching things on the screen.

With this definition in mind, there are just a few things we need to **touch on** here (😂).

First of all, it is a touch screen not a **jab** screen 🥊

Your author has seen many people literally pound their devices.

Touch screen devices resent the rough treatment (they will respond better to a light touch) and jabbing the screen puts unnecessary wear and tear on your joints.

Secondly, here are the gestures...

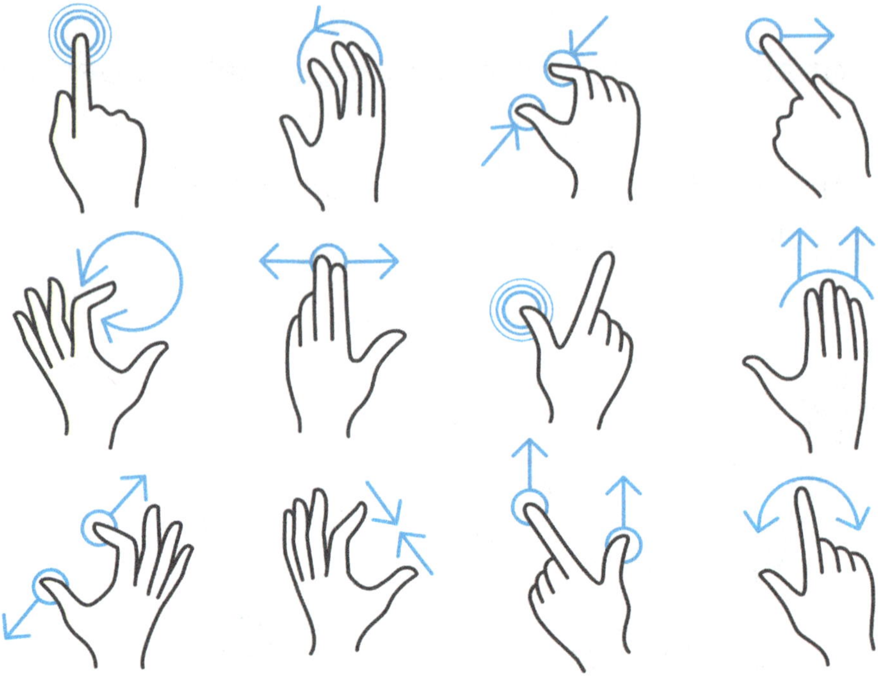

Yes, there are a lot of gestures (some of which you may never use).

But they are very easy to carry out and in time through **patience**, **persistence** and **practice** (i.e. the 3 P's), they will eventually become second nature.

What happens **after** you perform any of these gestures on a touch screen (or physical *trackpad*) depends on the **device** you are using, the **app** you are currently using (if any) and of course on our very good friend *context*.

Tap & hold

You can tap and hold over something (e.g. an *icon*) on a touch screen device to cause a *context menu* to pop-up:

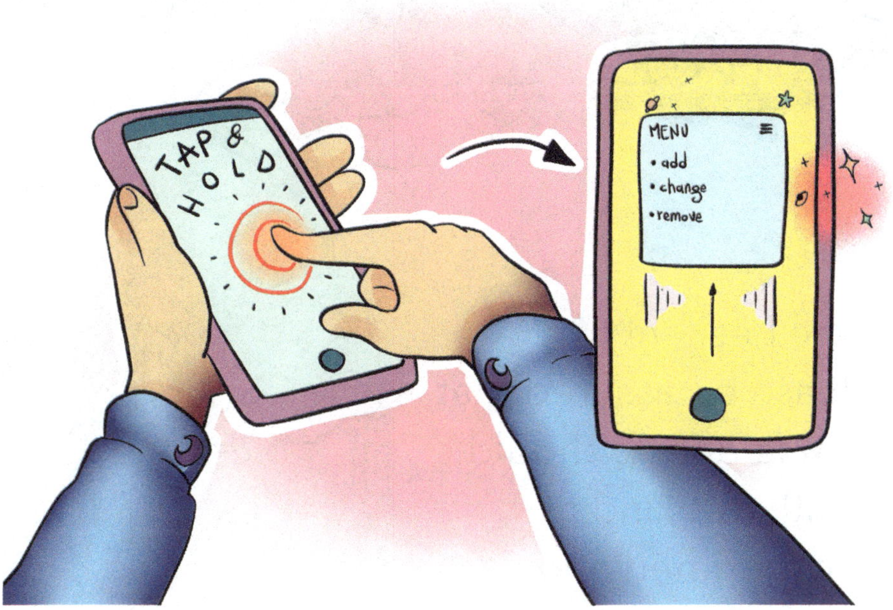

Whether (or not) a context menu appears – and the **selections** that will be on that menu – will depend on which *app* you are using and what the tip of your finger is currently over 🏝

Make a nice cup of tea and try out the other gestures on your device and see what happens.

If you are hesitant to do this, please be assured that no adverse effects should occur with this kind of casual experimentation.

Technology Basic #3: More than just typing, it's keyboarding!

Like our friend the mouse and our fingers (or stylus) on a touch screen, a keyboard is an **input device** – and needs to be used mindfully.

There are two types of keyboards:

Physical (wired and/or wireless connections)

Onscreen

Special function keys

Both types of keyboards work very much like a traditional typewriter when it comes to simple typing but they will also have some additional special function keys.

Some of these special function keys (such as sound volume and display brightness keys) may be intuitive while others will be a bit more abstract 🤔

Physical keyboards are usually designed to work with specific *operating systems* – which in turn will determine which special function keys are on a particular keyboard.

Pressing a keyboard's special function keys can have some unexpected results 🤨 if you are not cognizant of them.

So press them all and see what happens.

Please keep in mind that the keyboard's special function keys may do different things (if they do anything at all) depending on *context* 🏝️ (e.g. which *app* on your device you are currently using).

For example, pressing the "down arrow" (↓) key can scroll down a webpage.

(and conversely, pressing the up arrow key ↑ can scroll up a web page.)

Keyboard shortcuts

With practice, you can save time by giving the mouse a rest and using more keyboard shortcuts.

Some people have physical challenges which prevent them from using a physical mouse.

Therefore anything you can do with a mouse – such as open a *file* or start & close an *app* – you can **also** do with a physical keyboard.

Question: Which keyboard shortcuts can I use with my device?

Answer: That would depend on the device you are using and the keyboard you are using with that device.

You can *search the web* for keyboard shortcuts that you can use with your device and/or your device's *operating system.*

Keyboard shortcuts can lead to unexpected happenings 🤯

For example, "Wilby" would regularly use a library computer to type some very long email messages.

At some point, he would inevitably delete the text he had typed through the accidental use of the "Select All" keyboard shortcut.

Your author demonstrated to Wilby what would happen if the "Select All" keyboard shortcut had been accidentally performed in the midst of typing – **but** Wilby would not alter his method of keyboarding (that is he refused to look at the screen as he typed).

Therefore, Wilby continued to miss the *visual cues* which would have informed him that all of his text had been selected – and then subsequently deleted.

Technology obstinance often leads to lost time & lost work❣

Common keyboard shortcuts

Ctrl + C	→ copy selected	Ctrl + F	→ find
Ctrl + V	→ paste selected	Ctrl + S	→ save
Ctrl + X	→ cut selected	Ctrl + P	→ print
Ctrl + A	→ select all	Ctrl + Z	→ undo
Ctrl + R	→ refresh	Ctrl + Y	→ redo

Keyboard shortcuts will vary between operating systems.

Getting to Know Your Device...

Knowing what you are using is one of your technology foundation building blocks for some very good reasons (which we will explore as we progress through subsequent chapters).

Did you get your device as a gift?

Did you buy it new? Did somebody give you their old device? Do you use a public computer?

Regardless of what you are using or how you got it, the first step to becoming comfortable and productive with (whatever it is) is to say hello.

Even if you have already "met" your device, the beginning is always a great place to start – or to regroup

Just noodle around...

You can get acquainted (or reacquainted) with your device by just *noodling around*...

or by exploring...

Except we are not going to "explore" as much as "saunter around"...

We are certainly not going to joyride but simply do some casual sightseeing…

Three easy, no pressure objectives…

As you are noodling around your device, focus on…

1. seeing what your device has to offer
2. finding your device's "about" info
3. seeing how files can be stored on your device (if indeed files **can be** stored on your device) 🗄

Visual inspection

Sip your tea and look over the outer shell of your device of choice (or the one that you have to use under duress and protest 😒); give it the proverbial once or twice over and **see what's there**, take an unofficial inventory... ☕

Record your findings in your technology notebook

Well, bust my buttons!
–Doorman to the Emerald city

You should not "bust" the buttons on your device, but you may have to carefully feel around for them – because sometimes they blend right into the case.

Experiment! Press them – and double press them – to see what they do. Keep in mind that some button **behaviors** can be changed in the device's *Settings*.

The buttons you have on your device (and where they are and what they look like and what they will do) will depend on *which device* you have.

Ports of call

Ports can extend the capabilities of a device (if you ever need to do so).

Ports enable you to **physically connect** various things to your device.

Exactly **which** things you can connect will depend on the device itself.

Once you identify exactly *which device* you are using, you can *search the web* for info on its ports.

Who made this thing?

Look around your device for a brand name or logo – this is a **key piece of information** if you ever need *good help*.

A *visual inspection* will often reveal the **manufacturer** of the device but will not tell you how old the device is.

This notebook computer, for example, still looks modern if not downright futuristic (and still has many practical uses) but was actually released in the year 1999 and has been officially declared *obsolete* by its manufacturer:

TIP: If you do not recognize a logo (e.g.), you can ask a librarian.

Keep a knockin' but you can't come in?

LOGIN

Get some "good help" (Help for the Technology Shy Part II chapter 7) if you find yourself locked out of your device.

Unless your device is a *mystery gadget*, once you know its **manufacturer**, you should not have to go any farther than your local library for the customer support number.

Nothing should explode 🧯

Many of my technology shy friends have been hesitant to browse or noodle around with their devices out of concern that something will "break" or possibly explode.

Although nothing in life is 100% certain, a scenario like this is highly unlikely.

Think of this as tiptoeing through the tulips🌷🌷🌷🌷🌷 rather than treading into forbidden territory.

That being said, please take your time and observe & think **before** you click or tap 👀🤔

Please do not delete or change any *system files* on your device (although those files will be buried pretty deep on many devices and virtually inaccessible on other devices).

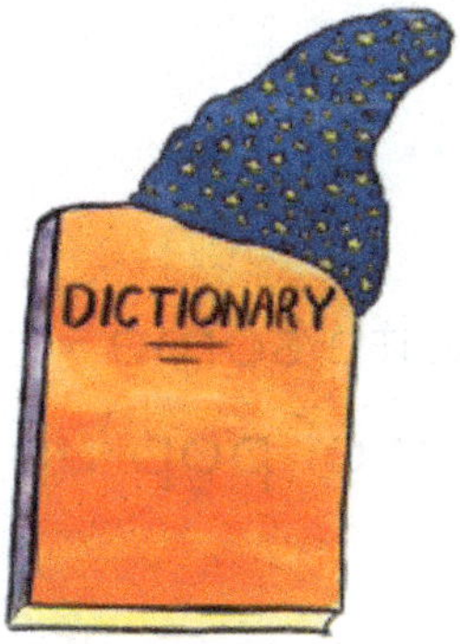

System files are those files that your device depends upon to operate properly

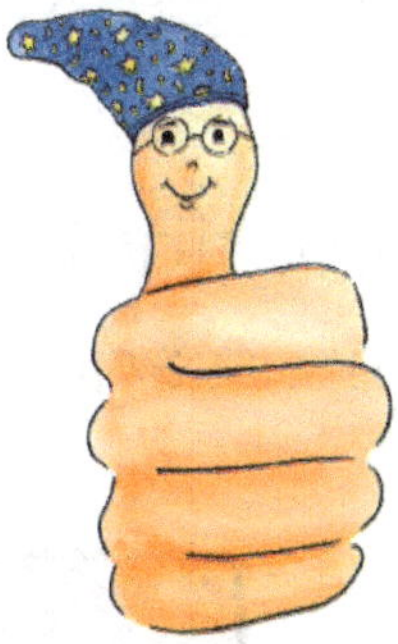

Please do not delete or change **anything** on your device unless you know exactly what you are deleting or changing.

Settings

As you are noodling around your device, you may come across something called "Settings" or "Preferences" or even "Control Panel."

You can use *Settings* (or whatever it is called on your device) to **customize** things on your device (including perhaps, making your screen text bigger).

Please also be wary of malware...

Just in case your device is infected with some form of *malware*, please be very **wary** of pop-up messages and fake notifications.

For right now, please do not do **anything** these messages tell you to do:

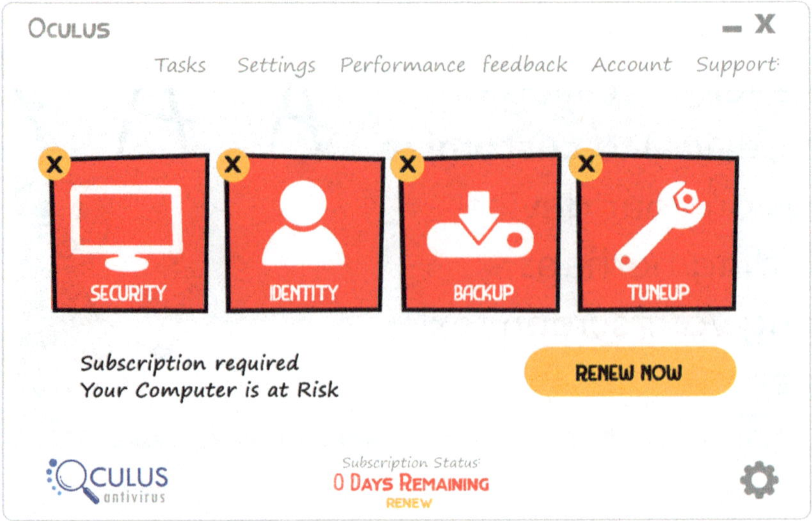

Look for *visual cues* (**X**) to close these pop-up messages.

sneaky adjective
\ ˈsnē-kē \

Definition of sneaky
: marked by stealth, furtiveness, or shiftiness

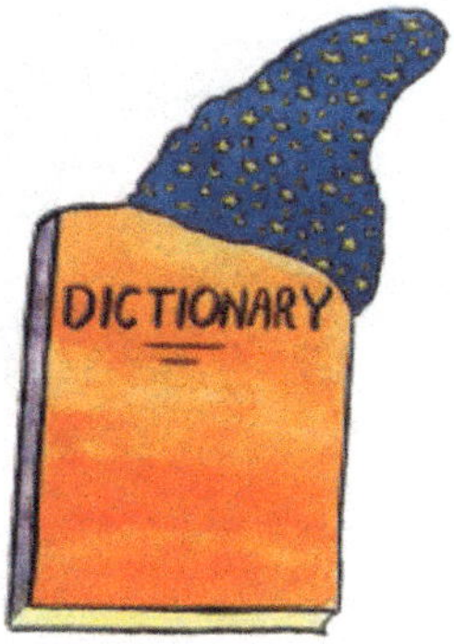

The close buttons on these "your device is at risk" windows (and others of that ilk) may be outright **shams** – they may not be close buttons at all 😨

The "close button" could instead be a **link** – taking you to a web page where you can buy whatever bogus "virus protection" they are peddling – or something even more nefarious 🦹

In any case, do not yield to any scare tactics! If necessary, completely power off your device and then get some *good help*.

Installation files

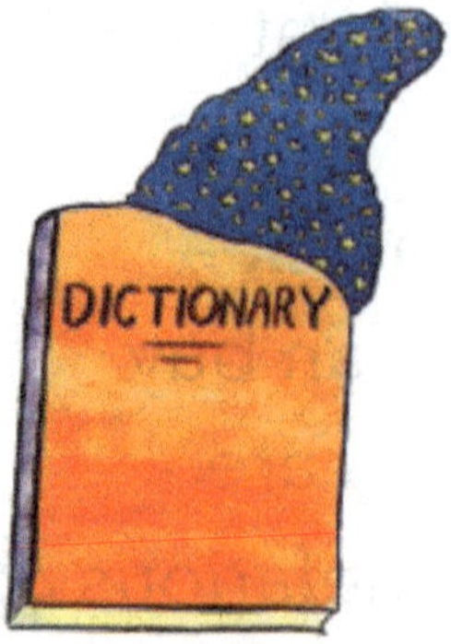

An installation program or installer is a program that installs files, such as applications, drivers, or other software, onto a computer (or other device).

In your web surfing travels, you could have inadvertently clicked or tapped a "Download" button...

And subsequently downloaded one or more *installation files* to your device...

Decline any offers to install anything onto your device unless you are within your device's *App Store* or are *very sure* of the source.

Assessing an interface (or simply seeing what's there)

You should calmly assess any screen you come across (or any screen that suddenly presents itself) **before** employing any of the technology basics.

The things that you see on a screen – and how they will look – will depend on the device and the *app* that you are currently using (if any) but the most common things will be *icons, buttons,* and *menus.*

There's no place like your home screen (or desktop)

A home screen is simply the first screen you will see after you power on and/or login to your device.

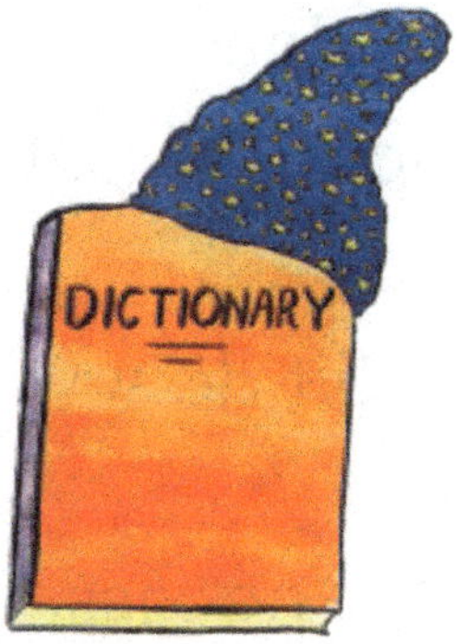

And like many things related to technology, desktops and home screens are highly customizable.

You can think of a home screen and a desktop as being launching pads. They are starting points for getting stuff done with a device ✅

When you are ready...

1. Make a cup of tea ☕ (fuzzy bunny slippers are optional but recommended 🐰 🐰).

2. Sit down with your device 💻🖥️ 📱

3. If necessary, power it on and login to it.

4. *Breathe*.

5. Look at your home screen and casually take in what you see (do not worry about understanding everything you see, the objective here is simple **familiarity**).

6. Start clicking or tapping things on your screen and see what happens (if anything).

7. Look for features and functionality (e.g. *search boxes*, *menus* and *buttons*).

There is no pressure here. There is no deadline. There is no sense of urgency. *Breathe* 🕊️ Enjoy the journey 🎒 Take notes if you like 📒

Look for navigation *breadcrumbs* to help find your way around...

Such as "go back," "go forward" & "go home"...

The **appearance** of navigation "breadcrumbs" will vary between devices. The important thing is to **look for them** and see what they do.

House of Baba Yaga

In Russian fairytales, Baba Yaga flies around in a mortar, wields a pestle, and dwells deep in the forest in a hut usually described as standing on chicken legs. Baba Yaga may help or hinder those that encounter or seek her out and may play a maternal role and has associations with forest wildlife.

(Included for fun and to emphasize that unfamiliar screen elements can look a bit unusual...)

Devices without screens

This example device doesn't have a screen at all:

But it does have *ports* and a remote... 🤔

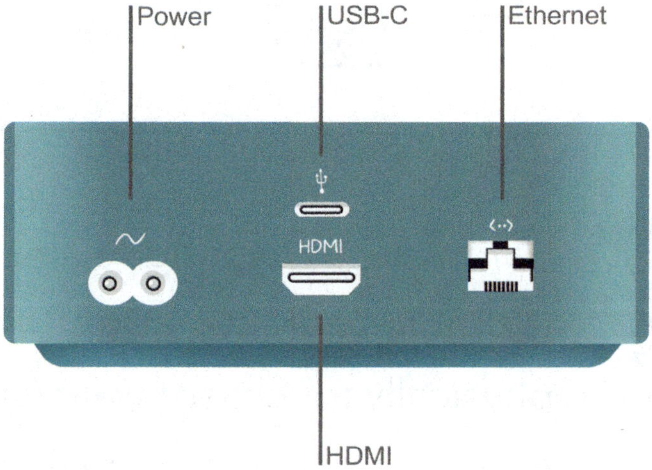

If such a device comes your way, inspect its outer case and look for a brand name and model.

Armed with this info, use an *internet-connected* device to *search the web* for...

- what the device is, exactly
- what you can use it for
- what other pieces of *hardware* you will need to use with it (such as a display)

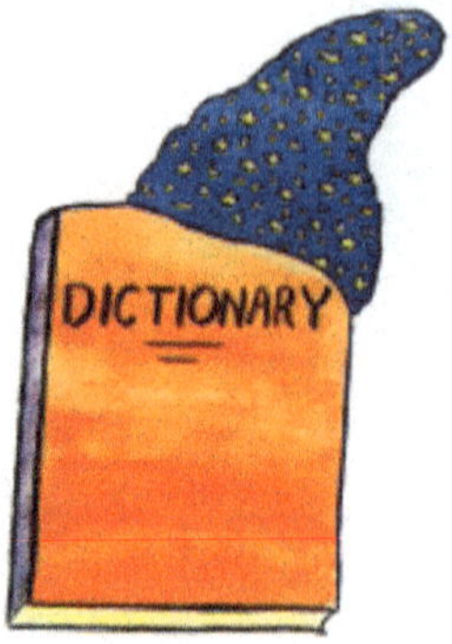

Hardware is anything related to technology that you can physically touch; *software* (such as an *app*) is anything that you cannot physically touch.

A note about OS updates...

If you **update** your device's *operating system* (or if your device updates its operating system on its own accord), things may change.

Appearances may change and **functionalities** may change.

An updated OS can be like a home interior **remodeling** in that the exterior will look just the same but inside it becomes a whole new house.

And, because your device's interior has been "remodeled," you may have to noodle around all over again...

OS updates (or "*system updates*") usually include security updates – which make a device less vulnerable to *malware*.

Does your device have a "file cabinet"?

Some of the buttons or icons or menu selections that you can click or tap (or even activate with *keyboard shortcuts*) may be the equivalent of opening a filing cabinet drawer 🗄️

If you come across them, casually take note. You will find it very beneficial to get an idea of how *files* are stored on your device.

You can **search directly** for *files* that are...

- stored on your device
- on an external storage device that is physically connected to your device
- in a *cloud account* that is connected to your device via an *app*

Searching for "About"

(Not as existential as it sounds 😉)

You can certainly find your device's "About" menu selection or button by noodling around 🍜 and following *breadcrumbs* 🍞 but you can also perhaps cut to the chase as it were and look around your screen for a search box or a search icon 🔍 and typing "about" into it...

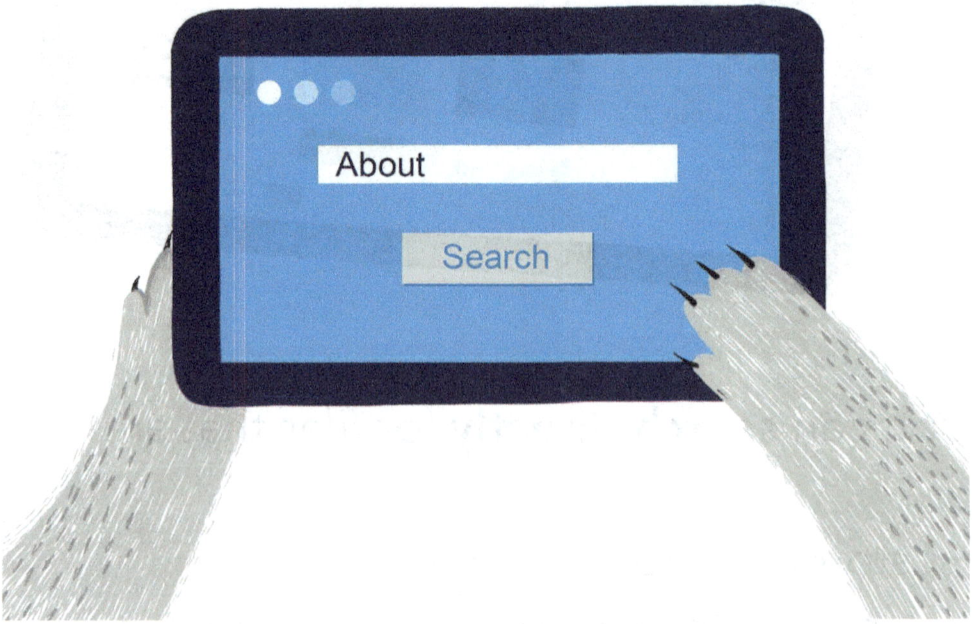

This might take you directly to your device's "About" info or you may have to follow some more *visual cues* to get to it 🍿

Can't find the search box?

At the time of this writing, a search box (or search *icon*) can usually be found **somewhere** on the *desktop* of a computer but are typically "hidden" (or tucked out of the way) on smartphones and tablets.

If you are using a smartphone or a tablet, finger gestures might reveal it...

If you know which model of device you have, you can *search the web* for particulars on using it (using another device if necessary).

And now without further adieu, your device's "About" screen may look something like this...

In your *Technology Notebook*...

1. Note the **name** of your device.

2. Record the "software version" (or *operating system*) your device is running.

3. Jot down your device's model number and serial number.

All roads lead to Rome from here. If there is something you want to do with your device, or if you are having trouble with your device, or if there is something you need to connect to your device, you will need to **first** positively ID it 🕵️‍♀️

All roads lead to rome.

Free from malware (aka "viruses")

Twenty-first century technologies have provided a *plethora* of opportunities for scam artists to take advantage of unsuspecting people.

The specifics may change but the **basic principles** of staying safe during technology use (e.g. proceeding mindfully, being **very** careful with your confidential information, and getting *good help* if you are not sure about something) will stay just the same.

Your stalwart & intrepid Help for the Technology Shy originator, public librarian emeritus and spiffy scribbler is **not** a security expert.

(Although he has very some good friends who are)

Your Tech Wizard Tour Guide
<u>does</u> know...

- when a device may have been compromised
- how to recognize fraud
- the importance of paying close attention to what you are doing
- how to *back up files*
- how to avoid *malware* (i.e. "viruses")
- how (and when) to *factory reset* a device 🍿

The information in this chapter is from your author's 25**+** years working with technology shy and **security conscious** adults, his own research and his actual, factual (and for the most part, practical) experiences with *devices*, *apps* and *cloud services* of every stripe.

Malware

In general, when people speak of "viruses" they are actually referring to **malware**.

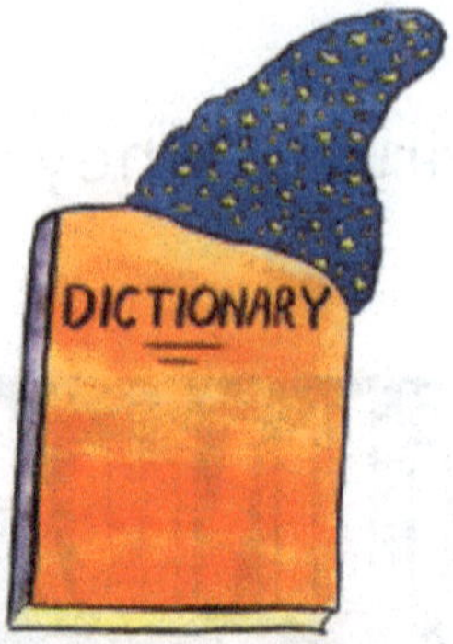

Malware is short for "malicious software" and its very specter can pose a significant obstacle to productive, stress-free technology use – and result in the theft of your personal information.

Types of malware

There are various forms of malware ranging from "adware"...

...to "spyware"...

...to "ransomware"...

... to trojan horses...

Some forms of malware are scarier than others but no form of malware on your device should be acceptable. It is all dangerous ⚠️

Malware is to devices what illnesses, injuries and infections are to the human body 🦠

But unlike the human body which can get better on its own without any form of treatment, a malware-infected device will **stay infected** until you do something about it.

Warning signs that your device might be infected with malware

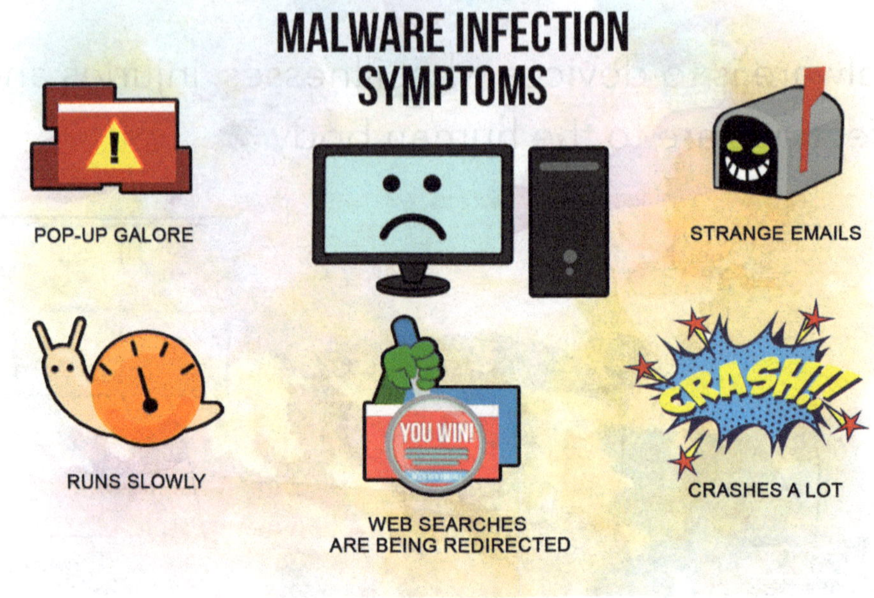

- your device runs very slowly (and may be nearly inoperable)

- your device "crashes" and needs to be restarted or powered off and powered back on again

- you get pop-up messages on your screen informing you that your device is "at risk" or has been "infected"

- your friends say they have gotten strange messages from you

- your web searches are being redirected to ads

These are just **warning signs**. They are not conclusive of malware infection.

For example, an *obsolete* device may run very slowly due to its age or condition.

How do devices get infected with malware?

1. through irresponsible technology use…

2. by opening bogus *email attachments*…

3. by clicking or tapping phony baloney "your device is at risk" pop-up windows...

4. by not paying **close attention** during technology use...

(For example, simply visiting a web page can cause a malicious *file* to automatically download in the background 🍿)

5. by trusting *nogoodniks*...

6. it comes **pre-installed** (wait, what? 😳)

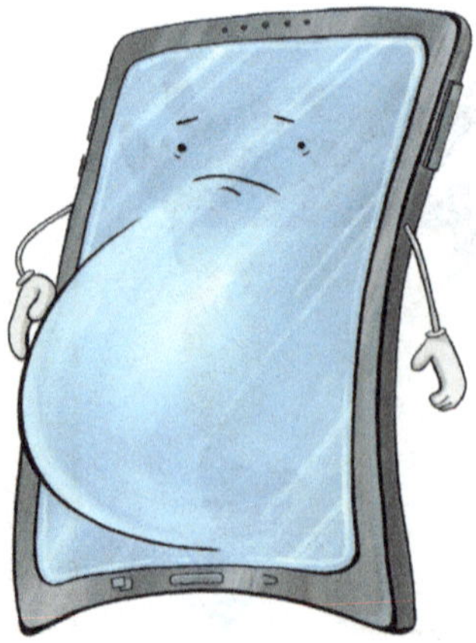

(We will get to "*bloatware*" shortly...)

Beeeee careful! How to avoid infecting your device with malware

Stay on the straight and narrow path when it comes to visiting websites (or "surfing the web").

It's a known fact, if you drive your car through the mud, it's going to get dirty.

If you drag your device through the "cyber-mud" so to speak, your chances of infecting it with malware will jump halfway to the moon 🌝

Practice "squeaky clean" web use

Simply avoid websites that offer online games, gambling, celebrity gossip, free movies & music, and coupons (as they are havens for malicious activities 😈).

Besides, real-world activities with positive people are much more fun 😃

Compare and contrast...

In your author's experience, many people do not know their devices are infected with malware because they do not know which *system notifications* and *web search results pages* are **legitimate** and which ones are not.

Compare how your device works with somebody else's device; see what their *system notifications* look like:

`If you are still not sure, get good help.`

Web searches that lead to a *plethora* of ads are tell-tale signs of malware on your device...

In general, in order to determine the possibility of malware on your device (and to be able to remove it), you need to know or find out...

- *which device* you are using
- which *operating system* is on your device
- what *system notifications* on your device are **supposed to** look like
- what *web searching* on your device is **supposed to** look like
- (if you are not sure, get *good help*)

Trojan horses

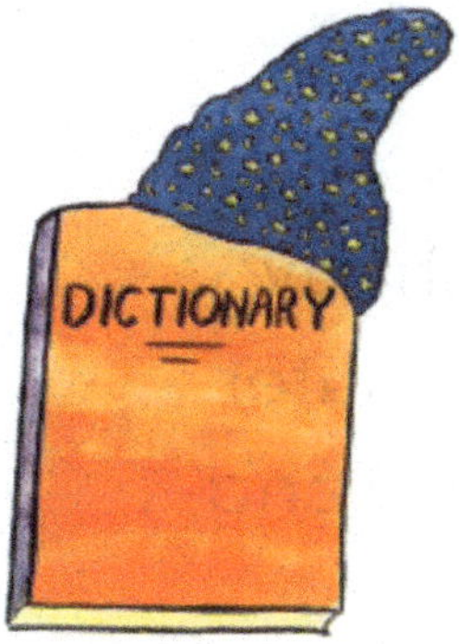

In the realm of technology, a Trojan horse, or Trojan, is any malware which misleads users of its true intent.

The term is derived from the Ancient Greek story of the deceptive Trojan Horse that led to the fall of the city of Troy.

Malware case study: tricked by a shady software company

Two of my technology-shyest friends unknowingly installed some a **very** shady application onto their notebook computer – which then started behaving strangely and erratically.

"Brandon" and "Edna" had no idea how the malware got onto their notebook computer but their patient and observant technology guide was able to quickly discern that they must have unknowingly **installed it themselves** due to the presence of *installation files* within their notebook's "Downloads" folder 🍿

In a nutshell...

- Brandon came across a "download" button on some webpage and clicked it (without knowing what he was downloading).

- And then at some point, Brandon saw the downloaded installation file and **opened it** by double-mouse clicking it ("knock knock") 🐭 🐭

- Brandon carefully (but not at all mindfully) followed the rogue application's installation procedure – which no doubt looked very legitimate (but was anything but).

Such a program can be the proverbial wolf in sheep clothing!

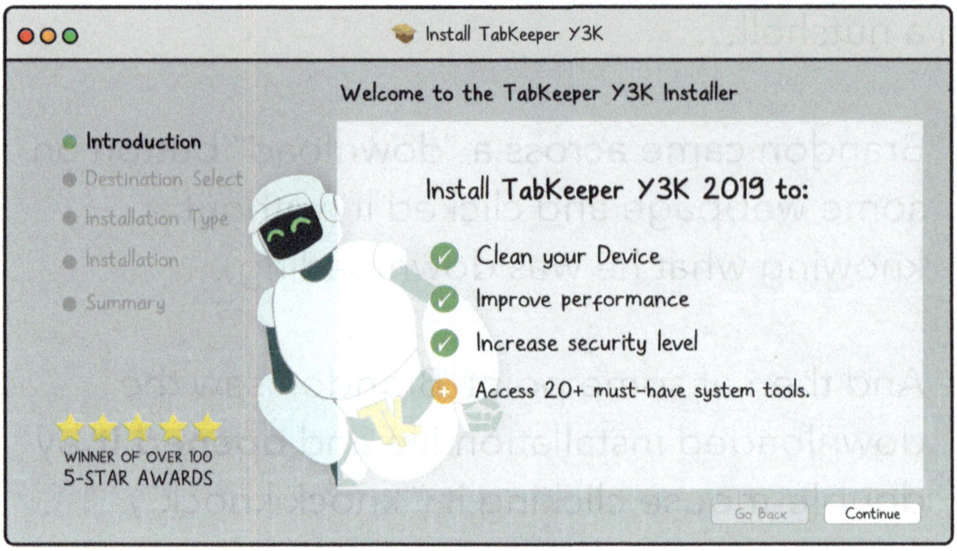

Once installed, this *rogue application* virtually hobbled their expensive notebook computer – effectively turning it into a high tech paperweight.

A product that **seems** very legitimate and beneficial can actually be a ruse, a ploy, a subterfuge, a scheme, or a contrivance of some sort designed to **trick you** into installing something horrible onto your device.

[Such applications can be useless, counterproductive and evil 😈 while still being legal ⚖️]

By all means, *noodle around your device* but please be the **very soul of caution** about downloading things to your device...

Stop, look & think before you click or tap

Tread carefully around set up and installation icons...

Knowledge is power
— School House Rocky

It is important to know what things are. If you are not sure what a file is, you can *search the web* for its *file name extension*:

Operating system updates

Scam scare tactics and "buy this *bloatware*" ads should not be confused with system update messages – that is messages that are actually from your device's operating system.

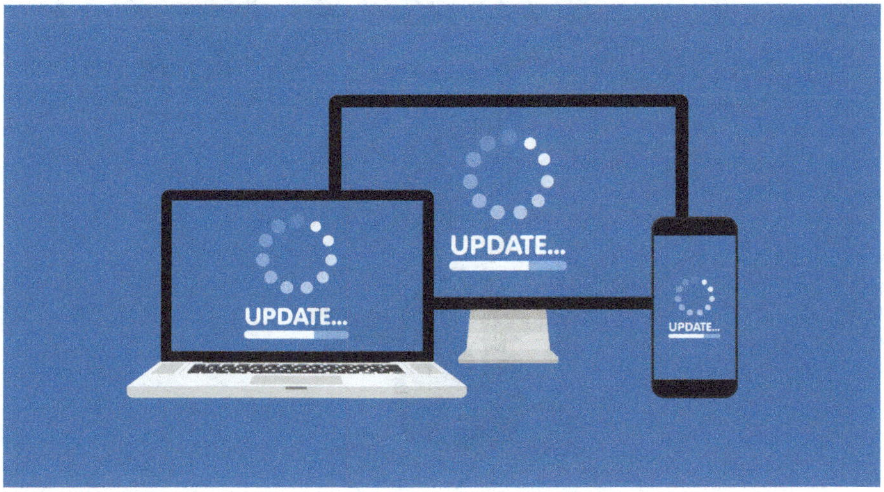

If you do not know what legitimate system update messages are supposed to look like on your device, you can...

- *Search the web.*
- Ask a friend who has the same device if he / she gets the same messages.
- Check with your device's (or operating system's) *support community*.
- Contact *customer support* (if possible).

Stop, look and think before you click or tap...

Some files on your device may be easy to identify (again it pays to be **visual** 👀) such as documents, pictures, videos and spreadsheets.

While other files may **not** be so self-evident…

The **appearance** of file *icons* will vary between devices.

If you are not sure what a file is, *search the web* for information on on it before you open it.

If you get info on a particular file and you **still** do not know what it is, then do not open it.

Stay in your device's own App Store

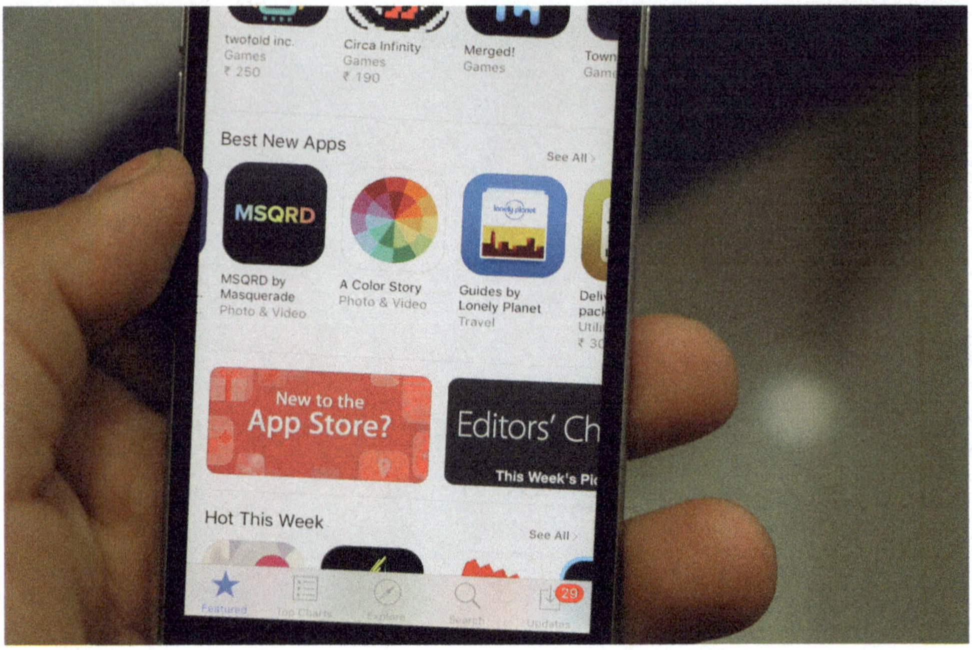

Yes, to be sure, you will find some clunky, junky apps in your device's App Store.

There may be also be apps in your device's App Store that you do not need because your device **will already have** the app(s) that you can use to accomplish a certain task or *objective* 🍿

But (although nothing in life is 💯% certain) none of the apps you will find within your device's App Store should be dangerous.

Rogue email attachments

My goodness, this is a biggie! This is how **a lot** of people infect their devices with malware.

Do not click or tap **any** email attachments unless you know and trust the sender of the message (we will discuss "spoofing" in the very next section).

And please do not click or tap any *installation files* (as only a *nogoodnik* or a *jackass* would attach one of those to an email message).

Question: What is an email attachment?

Answer: An email attachment is a *file* sent along with an email message. One or more files can be attached to any email message, and be sent along with it to the recipient. This is typically used as a simple method to share documents and images.

Question: How do you know if an email message has an attachment?

Answer: As always, look for *visual cues*

For example, a paper clip icon next to an email in your inbox would indicate a file has been attached to the message.

(Visual cues vary widely between apps, devices, operating systems and online services.)

“Spoofing”

The person you are communicating with – especially if they are asking you for money – may not be who you think they are.

Spoofing, in general, is a fraudulent or malicious practice in which communication is sent from an unknown source disguised as a source known to the receiver.

During the production of this very chapter, a message came in from a social media friend…

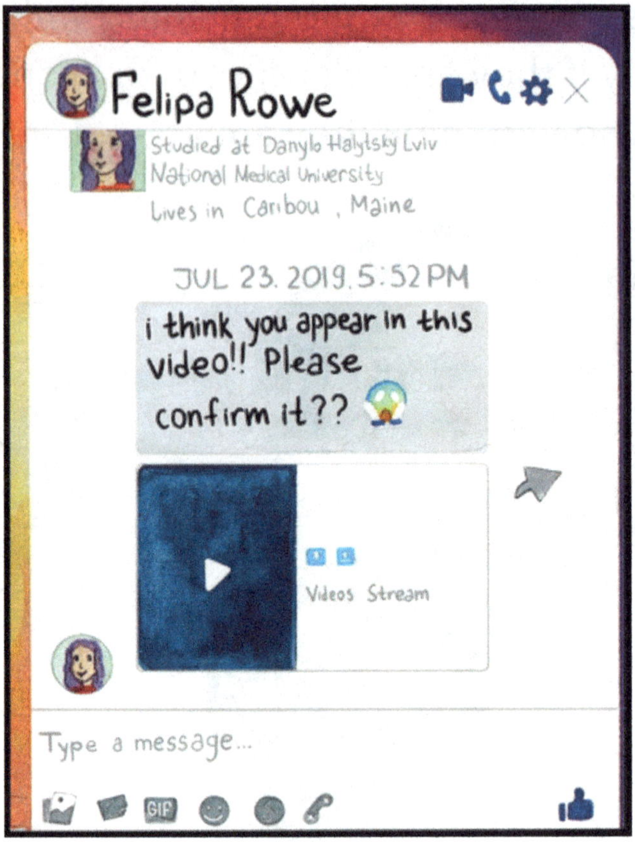

We looked at this message with both eyebrows raised…

… simply because the style and tone of the discourse felt **wrong**.

This particular social media friend always writes in complete sentences and uses proper punctuation and grammar.

It was also **out of character** for Felipa to send a link to a video and not to be cordial.

After contacting Felipa (outside the social media message system), she confirmed that she had not sent this message.

It was somebody **pretending** to be her.

As it turns out, Felipa's social media account had been accessed by a nefarious entity

If you get a strange or somehow sketchy message from a friend (or anybody else), contact them and ask for verification before you click or tap any links they send you.

Please do not fall for any blackmail scams

For example, if a website (or anything else) holds your web browser or your device itself "hostage" until you provide the information it wants or until you pay a ransom, simply power off your device.

Do **not** pay anything. Do **not** provide any of your confidential information. Do not engage, **period**.

(Even if the website is able to use your device's built-in camera to take your picture.)

A "ransomware" threat could result from visiting a fraudulent *website* – it does **not** necessarily mean that your device is infected with any form of malware.

The evil website could lock up your device but you might sort that by simply powering off your device and then turning it back on again (and then **never** visiting that website again).

If the ransomware persists after you power on your device (or restart it) – you will have some work to do...

1. Write down the specifics of the ransomware threat 📓

2. Bring your *technology notebook* (which will contain your device's model name and OS) and use a friend's device or a public device to do some research 🔍

The solution might be something as simple as "force quitting" your web browser and cleaning your web browser cache 🍿

3. Do not pay anything or buy anything.

4. *Good help* is not hard to find. Contact customer support for your device. See if there are **authorized** service providers in your area.

"Eric" received an email the morning after he visited an adult entertainment website. The writer of the email claimed that Eric's picture had been taken (using the built-in camera in Eric's notebook computer) while he was on that website and that picture would be sent to all of Eric's contacts if he did not send them money.

We did some research on this scam and thankfully, it turned out to be a bluff 🐿️

Bloatware

New computers, tablets, smartphones and wearables will have *apps* pre-installed on them that you can use for a variety of **useful** things (e.g. web browsing, email, note taking, et al.)

But some new devices will also come with equipped with "bloatware"...

Bloatware is a slang term for unnecessary programs that are **pre-installed** on some new devices.

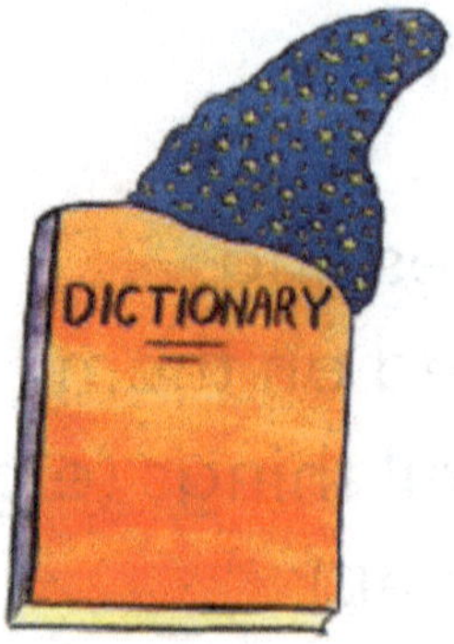

Bloatware is unnecessary software that will consume your device's resources (making it less than optimal for the things that you want to do with it).

Most bloatware isn't malicious **per se**, but it can be annoying (especially when it wants you to buy something) and can slow your device down to a crawl 🐌

Many of these programs are "lite" or limited trial versions designed to entice new users to buy or subscribe to the full-featured versions.

Many people install bloatware onto their devices because they do not know that the **functionality** that they seek (e.g. protection from *malware*) is **already present** on the device 😧

Question: If virus protection is **already** on the device, why is there a limited trial version for virus protection *software*?

Answer: It is the result of a mutually beneficial partnership between the manufacturer of the device itself and the creator of the antivirus product.

Due to sketchy (to use a kind word) sales tactics, people buy products & services they do not need (e.g. bloatware) which add additional layers of complexity to their technology use and often impair the functionality of their devices.

What to do about bloatware? And other forms of malware?

If you think your device has bloatware and/or malware, the simplest, err on the side of caution approach would be to *back up* all your important files & data, *factory reset* your device and then henceforward be visual and careful during technology use.

Factory reset

You can easily remove some forms of *malware* (especially *adware*) by...

1. identifying it (through observation) 📒
2. searching the web for a solution 🔎
3. carefully carrying out the steps in that solution 🍿
4. (getting *good help* if you need it)

But some forms of malware get so firmly entrenched, the only practical solution is a factory reset.

(You will not have to do anything like this!)

Factory reset is the term used to describe the removal of user data from an electronic device and restore it back to factory settings. A factory reset can be used to resolve some software issues associated with the device or simply to wipe all user data from a device.

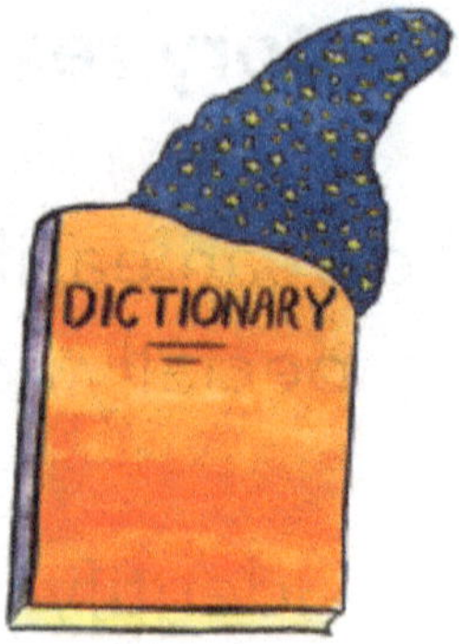

A factory reset will rid your device of malware (including *bloatware*) and make it "slim, trim & healthy."

And you will be very glad that all of your *passwords* – and what, exactly those passwords are for – were diligently recorded in your *technology notebook*.

How you can factory reset your device will depend on the device that you are using.

You can *search the web* for how to factory reset your device 🔎

You can telephone your device's customer support number (if available).

You can see if there are **authorized** service providers for your device in your area 📚

You can *watch* your favorite Help for the Technology Shy scribe back up and factory reset various devices.

Question: Which virus protection products does our favorite Tech Wizard use on his devices?

Answer: None.

(It is at this point that people look at your author as if he is crazy and believes in UFOs 👽🛸)

In your Technology Tour Guide's experience, **third party virus protection products** (that is those that are marketed & sold by a company other than the one who created your device's operating system) are unnecessary.

People usually (again, in your author's 25+ years experience) compromise their devices through a lack of mindfulness (e.g. willy-nilly mouse clicking and not paying attention to what is happening on their screens) and by falling for various and sundry scams

Question: Are some devices less susceptible to malware than others?

Answer: Yes. And those are the devices that your Tech Wizard wordsmith prefers to use.

Visit www.technologyshy.com/blog for current device (& workflow) recommendations.

Two examples of technology misuse...

Your Help for the Technology shy chronicler infected his work PC with malware through inattention and made a very expensive notebook computer virtually **inoperable** through sheer tomfoolery 😜

(The "tomfoolery" is in retrospect; hindsight being 20/20...)

(Tabby was not chagrinned, she is simply showing her solidarity.)

1. In 2016, your author downloaded an *app* to his library work computer from a website outside the device's *App Store* – without paying close attention to what was **actually happening** during the installation process:

It's worth noting here that the work computer did indeed have third party virus protection installed on it.

But the malware (a rogue browser toolbar) made the scene anyway 😒

2. Way back in 2005, your favorite Help for the Technology Shy chronicler purchased a new notebook computer for the tidy sum of $1,599.

A few weeks after the purchase, he read about an *app* in a reputable and well-known computer magazine that would customize his new notebook in all kinds of fun and exciting ways 🙄

Your author downloaded this app; installed it on his new notebook (unknowingly voiding the notebook's warranty); had fun with that customization app **until** his shiny new notebook suddenly went belly up.

It was toast.

So in order to avoid things like this from happening...

... we can add practice *technology minimalism*; use only what is necessary to our list of things to do (and not to do) in order to avoid malware, device destruction and other technology-related kerfuffles.

Off the Record, on the QT, & Very Hush-Hush 🤫

(Yep, We're Talking About Passwords!)

Let's face it, passwords are not the sexiest of technology-related topics.

But we cannot effectively use 21st century technologies without them 🔑

1970s Flashback

The first password your author had to remember was "rabbit" – which was necessary to gain access to the neighborhood clubhouse (and all of the board games, Kool-Aid™ and potato chips within).

That first password was easy to remember because Ellen (whose backyard the clubhouse was located in) was a rabbit fancier of the highest order.

Unfortunately, Ellen would **periodically** change the password for reasons of security (and it must be said for her own amusement).

If any of the club members failed to bring forward the **current** password, we could still (eventually) gain access to the clubhouse but Ellen would (sometimes literally) make us jump through hoops before she would tell us what the password was.

Nicknames (called "usernames" in the technology parlance of our times)

At some point, Ellen decided that club members should have nicknames.

So now in order to gain access to the clubhouse, you had to tell Ellen both your nickname **and** your password.

Real world nicknames are equivalent to technology usernames.

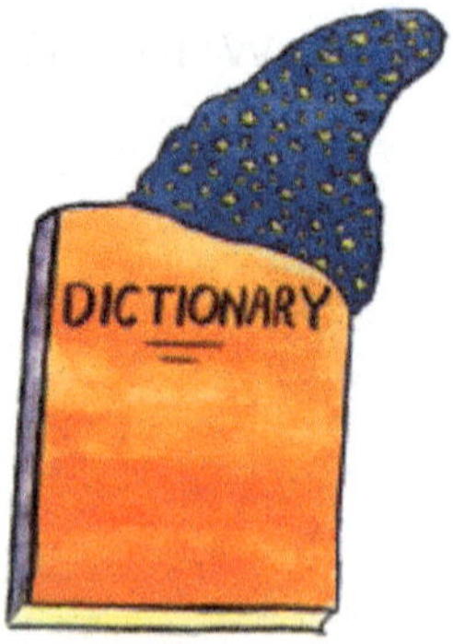

us•er•name
noun
1.an identification used by a person with access to a device, network, or online service.

A username can either be completely arbitrary or it might need to be your email address or something equally specific.

In either case, you need to keep track of what it is and what it is for.

The purposes and true nature of today's technology-related usernames, passwords and answers to security questions are much more **abstract** than the ones your author needed for access into the neighborhood clubhouse during the Watergate era – and therefore can be much more challenging to understand and keep track of.

What do we need usernames and passwords for?

1. to login into your device

2. to **administer** your device – passwords are often necessary to install new *apps*, for *system updates* 🔐 and to *factory reset* a device 🏭

3. to access your *cloud* account(s)

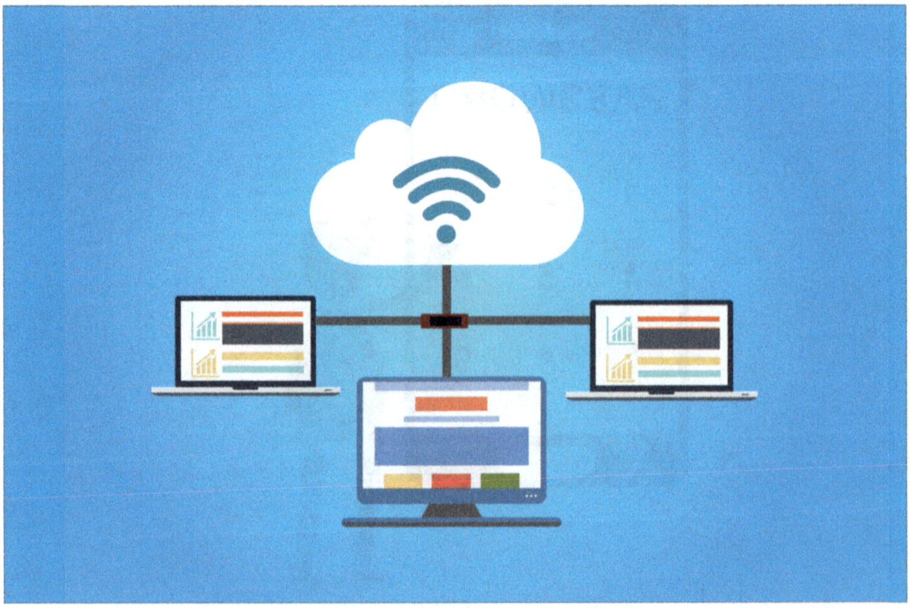

4. to access your various **online accounts** including email, banking, medical, gas & electric, insurance, music, shopping, home internet, mobile phone, your local library, social websites, and many, many more

In short, the number of usernames & passwords you need to keep track of can multiply and become **numerous** 🐰

An organizational system that you can trust...

You need a **simple** organizational system in place to keep track of your usernames & passwords.

And you need to keep this simple system **very neat**.

And you need to keep this neat, simple system **up-to-date**.

And you need to be able to **access** this system whenever it is called for.

The simplest way to keep track of all your usernames and passwords is with a designated technology notebook.

Creating a new username and password

Please do not try to rush through this admittedly ho-hum process without the necessary thought, understanding and attention.

Whenever you are called upon to create a new username and password...

1. Pause, come to a **full stop** ☕
2. **Understand** what this combo needs to be – and what it is **for** 🤔
3. Pick up a pencil ✏️ (or better yet some **colored** pencils)
4. Carefully **write down** your new username & password combo and what it is for
5. (Explanatory doodles are a bonus!)

This simple practice will not only save you much time (not to mention aggravation) but it will also increase your confidence as you will be **taking control** of this very important aspect of your technology use.

If you ever need to change or create a new username and/or a password, it will be of utmost importance to update your notebook – as in post haste, immediately, quick like a bunny, chop chop!!

Common technology notebook slip-ups

- not keeping the technology notebook up to date
- not keeping it neat and tidy
- not having it when it is needed
- leaving it unattended
- not writing down what a password (or username & password combo) is for
- having multiple, disorganized notebooks

Why have a strong password?

Because a weak password (e.g. a single word without any numbers and symbols) will leave you exceptionally **vulnerable** to all kinds of technology-related skullduggery & shenanigans.

But nothing in life is 💯% certain...

For example, one of Carol's online social accounts was accessed, vandalized and used to send obscene messages to her online friends resulting in serious embarrassment for her 🙀

Carol: "Psst! You are not going to believe what happened!"

Tech Wizard Mike: "Psst! Try me!"

This was especially perplexing, because Carol had diligently followed these traditional guidelines to create a "strong" password:

- at least 8 characters; the more characters, the better
- a mixture of both uppercase and lowercase letters
- a mixture of letters and numbers
- inclusion of at least one special character [such as: **! @ # ? * ^**]

Carol needed to take control of the situation by...

1. immediately changing her social account's password

2. contacting her social media friends to let them know that her account had been compromised

3. closely reviewing her bank & credit card statements for unauthorized purchases (which we should all be doing anyway as a matter of course)

Question: Why did this happen to Carol?

Answer: In your Tech Wizard friend's hands-on experience with such matters, getting "hacked into" can be like a car accident in which you were simply in the wrong place at the wrong time (figuratively speaking) 🐿

And this is probably what had occurred to Carol's online social account.

That being said, if your password for any of your online accounts is too simple (e.g. the name of your pet 🐶), you are incurring a **substantial risk** of being hacked into by automated "bots" that cruise the World Wide Web looking for accounts that are easily broken into.

Hooligans such as these generally do not infect physical but "live" (if that's the right word) to vandalize online accounts.

Use a passphrase instead?

According to the National Institute of Standards and Technology ("NIST"), **passphrases** are more effective than passwords. [1]

Passcode Example: Rocket Mary Stars Pencil Coffee

---

[1] Source: https://www.us-cert.gov/

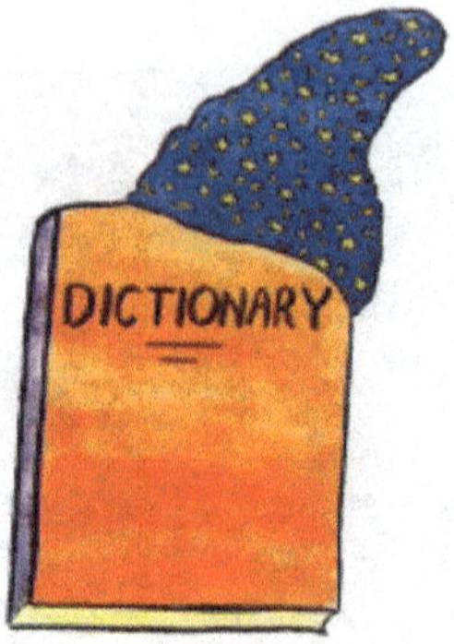

A passphrase is a sequence of words or other text used to control access to a computer system, program or data. A passphrase is similar to a password in usage, but is generally longer for added security.

Common password (or passphrase) slip-ups

- not recording them immediately
- rushing through the process of creating them
- creating passwords or passphrases that are too short
- using the **same** passwords or passphrases for multiple accounts

Save your password?

Your device (or more technically an *app* on your device) may offer to **save** your passwords – usually after you have typed one into a web form.

Always say "never" if you are using a device that you do not personally own

NOTE: "Password" will now be a generic catch-all term for confidential information in general including credit cards.

Question: Is it **safe** to let my device save my passwords?

Answer: **If** you are the only one that has access to your device and **if** your device is free from *malware* then it will **most likely** (although nothing in life is 💯**%** certain) be okay to save your passwords on your device.

TIP: Explore your device's login security options.

Common save passwords on device slip-ups

- not *backing up* your passwords (a back up would be necessary if your device is stolen or if you sell it or it becomes inoperable)

- using another person's device or a public device (while **depending** on saved passwords)

- not knowing **where** the passwords are being stored (e.g. within the *web browser* app or on the device itself) 🍿

- saving your passwords on a device that is infected with *malware* which could also lead to your online accounts being compromised because your passwords (and other private data) could be transmitted to nefarious third parties 🦹

Security measures such as Touch ID and facial recognition may have taken the place of passwords by the time you read this but it will always be of great value to **plan ahead** – to make sure you have what you need, when you need it.

Even if you are saving your passwords onto your device, write them down in your technology notebook.
Review **all** the information in your *technology notebook* on a regular basis.

Please do not fall into the "Fixodent® and Forget It" mindset when it comes to any aspects of your technology use.

Third party password managers

Many of my technology shy friends have asked about password managers.

Third party password managers are *apps* made by a company **other** than the one who created your device's *operating system*.

A password manager app assists in generating and retrieving complex passwords, potentially storing such passwords in an encrypted database or calculating them on demand 😳

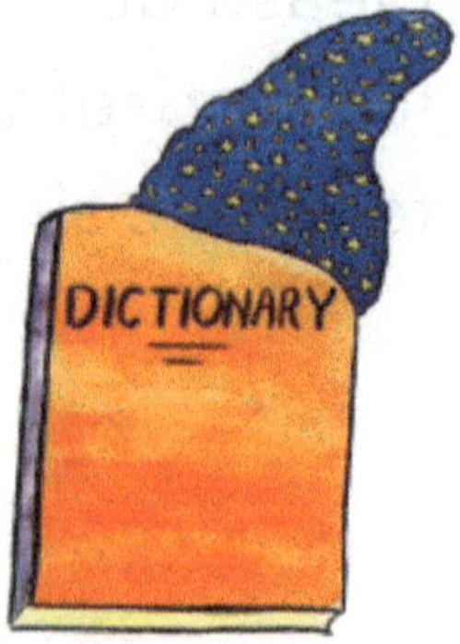

In short, they will manage your passwords for you and you just need to remember a single password

.

Question: Does you **personally** use a third party password manager app?

Answer: No, at the time of this writing, your Tech Wizard friend's computers, tablets, smartphone and wearable are all from the same manufacturer and are connected to each other through that manufacturer's *cloud* services.

Question: Do you **recommend** the use of password manager apps in general? Especially for people who are using multiple devices made by different manufacturers? 📱💻

Answer: In your author's experience, people should definitely **not** use password manager apps **until** they feel they are in complete control of their technology use.

A note from the Help for the Technology Shy "Tough Love" Department...

A password manager should be a well-understood convenience.

It should not take the place of having to **think** 🤔; it should not take the place of **understanding** what passwords are for – as this would almost certainly lead to additional confusion, displeasure, pique, gnashing of teeth, pulling of hair, rending of garments, etc.

A technology *notebook* / diary / journal will help you **take control** of your technology use and as a result make you feel much less anxious about it.

This book and "Tabby's Guide to Thoughtful Tea Drinking" make for great bedside reading!

Help for the Technology Shy Part II

Now, with your emotional & mental "bedrock" in place...

And your technology foundation firmly placed on top of that bedrock...

We can now focus on **getting stuff done**.

Available in print and digital versions, Help for the Technology Shy Part II can be ordered from Amazon or from www.technologyshy.com.

Ask your local librarian and/or bookseller to order Help for the Technology Shy books from Ingram.

Index

(**NOTE**: This index includes Help for the Technology Shy Part II.)

A

B

C

F

G

H

J

K

M

N

O

P

R

S

T

W

The fabulous illustrators for this book (and Help for the Technology Shy Part II) were engaged through upwork.com.

The case studies and other examples in this book were based on actual tech help sessions but the names and appearances of the people were changed.

Made in the USA
Monee, IL
02 November 2026

67381631R00155